STREAMFLOW CHARACTERISTICS

OTHER TITLES IN THIS SERIES

1 G. BUGLIARELLO AND F. GUNTER
COMPUTER SYSTEMS AND WATER RESOURCES

2 H.L. GOLTERMAN
PHYSIOLOGICAL LIMNOLOGY

3 Y.Y. HAIMES, W.A. HALL AND H.T. FREEDMAN
MULTIOBJECTIVE OPTIMIZATION IN WATER RESOURCES SYSTEMS: THE SURROGATE WORTH TRADE-OFF-METHOD

4 J.J. FRIED
GROUNDWATER POLLUTION

5 N. RAJARATNAM
TURBULENT JETS

6 D. STEPHENSON
PIPELINE DESIGN FOR WATER ENGINEERS

7 V. HÁLEK AND J. ŠVEC
GROUNDWATER HYDRAULICS

8 J. BALEK
HYDROLOGY AND WATER RESOURCES IN TROPICAL AFRICA

9 T.A. McMAHON AND R.G. MEIN
RESERVOIR CAPACITY AND YIELD

10 G. KOVÁCS
SEEPAGE HYDRAULICS

11 W.H. GRAF AND C.H. MORTIMER (EDITORS)
HYDRODYNAMICS OF LAKES: PROCEEDINGS OF A SYMPOSIUM 12–13 OCTOBER 1978, LAUSANNE, SWITZERLAND

12 W. BACK AND D.A. STEPHENSON (EDITORS)
CONTEMPORARY HYDROGEOLOGY: THE GEORGE BURKE MAXEY MEMORIAL VOLUME

13 M.A. MARIÑO AND J.N. LUTHIN
SEEPAGE AND GROUNDWATER

14 D. STEPHENSON
STORMWATER HYDROLOGY AND DRAINAGE

15 D. STEPHENSON
PIPLELINE DESIGN FOR WATER ENGINEERS
(completely revised edition of Vol. 6 in the series)

16 W. BACK AND R. LÉTOLLE (EDITORS)
SYMPOSIUM ON GEOCHEMISTRY OF GROUNDWATER

17 A.H. EL-SHAARAWI (EDITOR) IN COLLABORATION WITH S.R. ESTERBY
TIME SERIES METHODS IN HYDROSCIENCES

18 J. BALEK
HYDROLOGY AND WATER RESOURCES IN TROPICAL REGIONS

19 D. STEPHENSON
PIPEFLOW ANALYSIS

20 I. ZAVOIANU
MORPHOMETRY OF DRAINAGE BASINS

21 M.M.A. SHAHIN
HYDROLOGY OF THE NILE BASIN

STREAMFLOW CHARACTERISTICS

H. C. RIGGS

3415 Executive Avenue, Falls Church, VA 22042, U.S.A.

ELSEVIER

Amsterdam — Oxford — New York — Tokyo 1985

ELSEVIER SCIENCE PUBLISHERS B.V.
Molenwerf 1
P.O. Box 211, 1000 AE Amsterdam, The Netherlands

Distributors for the United States and Canada:

ELSEVIER SCIENCE PUBLISHING COMPANY INC.
52, Vanderbilt Avenue
New York, NY 10017

ISBN 0-444-42480-6 (Vol. 22)
ISBN 0-444-41669-2 (Series)

Printed in The Netherlands

PREFACE

Streamflow is highly variable. A flood discharge may be thousands of times greater than the discharge during a drought. Such flow regimes can be described by various statistics collectively referred to as streamflow characteristics. These characteristics provide information needed in the design of structures built in or along stream channels, for evaluating flood hazards, and for defining the available water supply. Various methods may be used to compute flow characteristics; the appropriate ones for a particular stream depend on its flow regime and on the amount and type of data available. Although many methods of defining flow characteristics have been proposed, and are being used, descriptions and evaluations of these are widely scattered in the literature. This book brings together some of the more useful methods - ones that are simple, practical, and require only commonly available or readily obtainable data. These methods produce results comparable in accuracy with those from more sophisticated methods for many problems.

An analyst needs more than descriptions of methods. He needs to understand the hydrology in the region studied and the principles of the statistical techniques used, and to have some feel for the reliability of his data. Discussions of the effects of climate, geology, and topography on the streamflow regime, of statistical principles, and of methods for collecting hydrologic data respond to these needs.

The analyst should know how his interpretations will be used. Some of the principal applications of streamflow characteristics to water-related problems of engineering and management are included for that purpose. Also included is information on the changes to be expected in streamflow characteristics because of natural or man-made changes on the land.

The material in this book was selected for its value to practicing hydrologists; the book is intended to be both a manual and a source of references. Engineers should find some parts useful for appraising the reliability of hydrologic information on which their designs will be based. Environmentalists may find the interpretations of various streamflow characteristics helpful.

This is not a hydrology book in the usual sense. Some important elements of the hydrologic cycle are not covered in detail because adequate data on them are often difficult to obtain; these elements are not used in the techniques given here. Likewise, certain well-known techniques are not included because they are too time-consuming for common use. Ground water is included only to the extent needed to properly evaluate the surface-water hydrology. Water quality is discussed briefly.

The material has been drawn from many sources, the principal ones being the literature of government agencies and professional societies. I am particularly indebted to the U.S. Geological Survey which provided me the opportunity to participate in many hydrologic investigations, and for use of examples and explanatory material from their many publications on hydrology. I also acknowledge the permission of Pergamon Press to include parts of my articles in climatic factors of runoff (Chapter 2) and in snowmelt runoff (Chapter 12).

CONTENTS

Chapter 1

INTRODUCTION

1.1 GENERAL

Streamflow changes continually in response to weather and the modifying effect of the land. But the climate over a drainage basin follows a pattern throughout each year and this results in a yearly cycle in streamflow. On this cycle are imposed the variations of flow because of the particular weather in the preceding days or months. Both the climate and the land vary from stream to stream and thus the flows also differ.

Hydrologists explain the movement of water on the earth by the hydrologic cycle, an extremely generalized concept that begins with evaporation of moisture from the oceans and transport of that moisture in cloud masses over land where it is precipitated. The precipitation, if rain, may be disposed of by runoff on the land surface, infiltration into the soil, and by evaporation. The infiltrated water may remain as soil moisture to be later evaporated or transpired by plants, or some of it may move down to a water table from which it may reappear as streamflow. Evaporation and transpiration take an additional toll from streamflow before that water returns to the ocean. Obviously the transformation of precipitation into streamflow in a particular basin is extremely complicated because of the many variables of weather, geology, and topography. Although the land phase of the hydrologic cycle is well understood, the input and the many relevant land characteristics cannot be measured accurately enough to define streamflow completely deterministically. Chapter 2 describes how various measurable factors affect the flow regime. General principles are explained in texts on hydrology such as one by Linsley, Kohler, and Paulhus (1982).

Quantitative hydrology is a rather recent science. Streamflow measurements were not widely available until after 1888 when the U.S. Geological Survey began systematic gaging. Subsequent demands for irrigation water, hydropower, flood control, municipal and industrial water supplies, and soil conservation led to major activity in hydrologic analysis. Fuller and Hazen had studied the frequency distributions of floods by 1914 and flow-duration curves and methods of estimating flows of ungaged streams were available at that time. Jarvis (1936) and Hoyt (1936) give brief summaries of developments up the the mid 1930s. Many of the developments since then resulted from greater use of statistics.

The availability of digital computers further expanded the capability to utilize large amounts of data in analyses that were not previously feasible. The improved understanding of ground-water movement also has contributed to progress in surface-water hydrology; surface and ground water are intimately related in most regions.

Streamflow characteristics are quantitative measures of the magnitudes and variabilities of daily, monthly, and annual means and of flow extremes. They can be computed from streamflow records or by indirect means where no records are available; these computed values are of course only estimates of the long-term values.

Streamflow characteristics are used in a wide variety of problems such as defining available flow with and without storage; forecasting floods, droughts, and snowmelt runoff; designing bridge waterway openings; planning protection from floods; and establishing regulatory requirements. Probable changes in flow characteristics resulting either from natural or artificial modifications of a basin are needed to evaluate the impact of such modifications.

The material presented here is directed to hydrologists who are faced with practical problems and to engineers and planners who use hydrologic data and would profit from knowing how it was derived and how it should be interpreted.

This book provides selected techniques and the other information needed to estimate streamflow characteristics and to evaluate their reliability. The literature on hydrology and hydrologic analysis is extensive. Any collection of techniques in a book of reasonable length would not be complete. The techniques selected for inclusion in this book are limited to ones that are reasonably simple to define and that require only commonly available data. More complex methods are mentioned and referenced.

Any description of a hydrologic technique is incomplete without some means of evaluating the reliability of the result. Such evaluations should be based on an understanding of the hydrology of the basin or region, of statistical principles, and of the reliability of the data used. The chapters on Factors Affecting Streamflow, Collection of Hydrologic Data, Relation of Ground Water to Surface Water, and Statistics are included to provide that understanding. Additionally, throughout the book evaluations and limitations of the various techniques are discussed. This information should help the reader to select the most appropriate technique for the particular application as well as to make a (subjective) estimate of the reliability of the result.

Many statistical techniques applicable to hydrology can be applied without knowledge of statistical theory. For example, computer programs for frequency analysis and multiple regression can be used, and often are, without knowledge of either statistics or hydrology. But an analyst with no statistical background may make assumptions of which he is not aware and he may incorrectly interpret his results. The material in the chapter on statistics is directed to those with little training in that field; it concentrates on theory (non-mathematical) rather than on computational details.

By emphasizing applied hydrology in this book, the coverage of practical subjects is broader and the analyses are described in more detail than is usual in one source. On the other hand the hydrologic subjects are limited to those useful in applying the techniques presented. Consequently, several subjects commonly included in hydrology texts do not appear here. Among them are the unit-hydrograph method, infiltration, characterizing precipitation, flow routing, channel hydraulics, evapotranspiration, computer modeling, and estimation of flow characteristics from precipitation in arid and semiarid regions.

1.2 PROBABILITY AND RECURRENCE INTERVAL

For many years engineers have described the frequency characteristics of annual events in terms of recurrence interval, the reciprocal of probability. For example, the annual flood with a probability of 0.01 of being exceeded in any one year is commonly called the 100-year flood. Either way of designating that flood is correct but because laymen often misinterpret the meaning of a flood of a given recurrence interval, the U.S. Water Resources Council recommended that the preferred descriptor be probability. Recurrence interval is generally used in this book because recurrence interval was used in most of the analyses selected as examples. No preference for recurrence interval should be implied.

1.3 UNITS

Some English-speaking nations use the British or Imperial system of units although in surveying and in hydrology tenths and hundredths of feet are used instead of inches. International journals require use of metric or Systeme International d'Unites (SI units) which have been adopted by nearly all scientists. A complete and abrupt change to the metric system in hydrology is not feasible in the United States because many existing laws, regulations, records, and standards are based on the British system. For demonstrating principles and explaining techniques either system of units is adequate. The problems and examples given in this book are in the units in which they were originally reported. Results can easily be converted from one system of units to the other.

1.4 CONVERSION FACTORS

1 inch = 25.4 millimeters

1 foot = 12 inches = 304.8 millimeters

1 mile = 5280 feet = 1.609 kilometers

1 millimeter = 0.039 inches

1 meter = 39.37 inches = 3.281 feet

1 kilometer = 0.621 miles

1 acre = 43,560 square feet

1 square mile = 640 acres

1 square meter = 10.76 square feet

1 hectare = 2.471 acres

1 square kilometer = 100 hectares = 0.386 square miles

1 cubic foot = 7.4805 U.S. gallons = 0.028 cubic meters

1 U.S. gallon = 3.785 liters

1 acre-foot = 43,560 cubic feet = 1233.5 cubic meters

1 cubic meter = 35.31 cubic feet

1 million cubic meters = 810.7 acre-feet

1 cfs (cubic foot per second) = 0.028 cms (cubic meters per second)

1 cfs-day = 0.0864 million cubic feet
= 1.983471 acre feet
= 0.6463 million gallons

1 cfs-year = 723.97 acre feet (for 365 days)

1 cfs per square mile
for one day = 0.03719 inches depth on the drainage basin

Annual runoff in inches = 13.58 times annual mean, in cfs per square mile

1 pound = 0.4536 kilograms

1 gallon = 8.33 pounds (water)

1 cubic foot = 62.4 pounds (water)

1 kilogram = 2.205 pounds

1 foot per second = 0.682 miles per hour

REFERENCES

Hoyt, W.G., 1936, Studies of Relations of Rainfall and Runoff in the United States: U.S. Geological Survey Water-Supply Paper 772, 301 p.

Jarvis, C.S., 1936, Floods in the United States, Magnitude and Frequency: U.S. Geological Survey Water-Supply Paper 771, 497 p.

Linsley, R.K., Kohler, M.A., and Paulhus, J.L.H., 1982, Hydrology for Engineers, Third Edition: New York, McGraw-Hill, 496 p.

Chapter 2

FACTORS AFFECTING STREAMFLOW

2.1 INTRODUCTION

The process by which precipitation becomes streamflow is well known in a general way. But for a specific drainage basin it is not possible to quantify all the relevant variables and to describe the paths that the water takes at various times. Nevertheless some knowledge of the probable processes in a basin is essential if realistic estimates of streamflow characteristics in that basin are to be obtained. Climate, geology, and topography are the principal basin characteristics affecting streamflow. Unusual streamflow patterns usually can be explained qualitatively by some feature or combination of features of these. This chapter provides such explanations using examples of flows from various regimes. More information on hydrologic characteristics throughout the world is given by L'vovich (1980). The following section, 2.2, is from Riggs (1980) by permission of Pergamon Press.

2.2 CLIMATIC FACTORS

The climatic factors of precipitation, temperature, sunshine, humidity, and wind all affect stream runoff to some extent, but only precipitation and temperature account for major differences among runoff regimes in regions of similar geology and topography. Precipitation, the source of runoff, is disposed of in several ways, some is evaporated from the surfaces on which it falls, some infiltrates into the soil, and some moves directly to stream channels. Of the water that infiltrates the soil, some will be removed later by evaporation and transpiration and some may reach a ground-water body which maintains stream runoff between storms. Temperature, in conjunction with sunshine and wind, determines the evapotranspiration losses. In temperate and tropical regions of low annual precipitation, evapotranspiration losses may nearly equal the precipitation. At the other extreme, evapotranspiration may be as little as 10 percent of precipitation in regions of high rainfall. Below-freezing temperatures affect runoff by delaying runoff from snowfall and by modifying the winter runoff because of ice formation.

2.2.1 Effects of the precipitation regime

The response of Shetucket River to an intense storm rainfall associated with a West Indian hurricane is shown in Figure 2.1. The extremely high peak discharge is followed by a high base flow resulting from the recharge of ground water in the drainage basin.

The magnitude of a flood peak is not only a function of the rainfall magnitude and duration, but also of the areal coverage of the storm relative to the

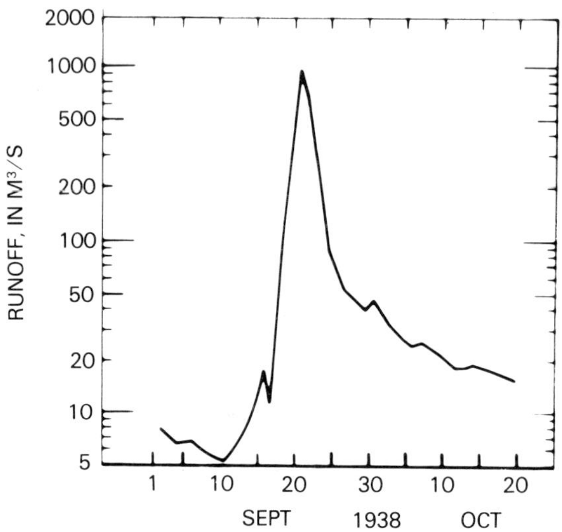

Fig. 2.1. Runoff from intense rainfall, Shetucket River, Connecticut.

drainage basin of the stream. Thus the character of storms in a region determines the relative magnitudes of unusual floods for various drainage area sizes. In some arid and semiarid regions thunderstorms produce exceptional flood peak discharges per unit area for the smaller drainage areas. Thunderstorms also produce record floods from small basins in humid regions, but those regions are also subject to more widespread storms which produce the major floods from larger basins. The difference in flood potential between a semiarid and a humid region is shown by plots of maximum known flood-peak discharges (Crippen and Bue, 1977) against drainage area for semiarid Arizona and for humid Virginia (Fig. 2.2). The lesser flood potential from large basins in Arizona is due principally to the limited areal extent of major storms.

In basins where snow does not accumulate, and where basin characteristics are not unusual, the seasonal distribution of runoff closely follows the seasonal distribution of precipitation. Figure 2.3 compares the monthly precipitation to the resulting runoff for Seti Khola, Nepal, in a region of high rainfall. Typical distribution of runoff in a climate with wet winters and dry summers also follows the precipitation closely so long as temperature is generally above freezing (Fig. 2.4).

Seasonal variability of large streams tends to be less than for small ones because large ones may have sources in different climates so that the cumulative

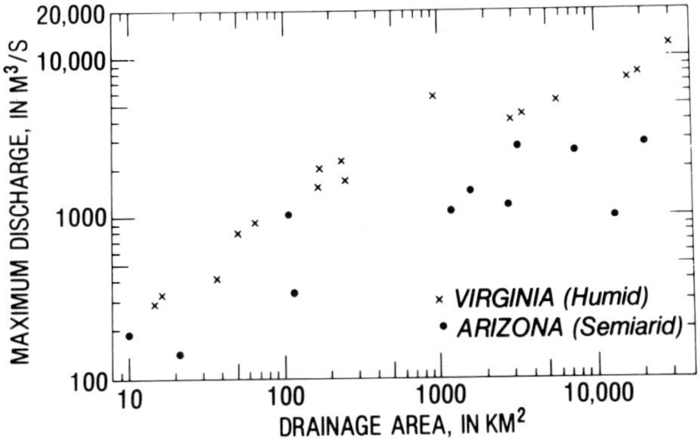

Fig. 2.2 Maximum known floods in a humid and a semiarid area.

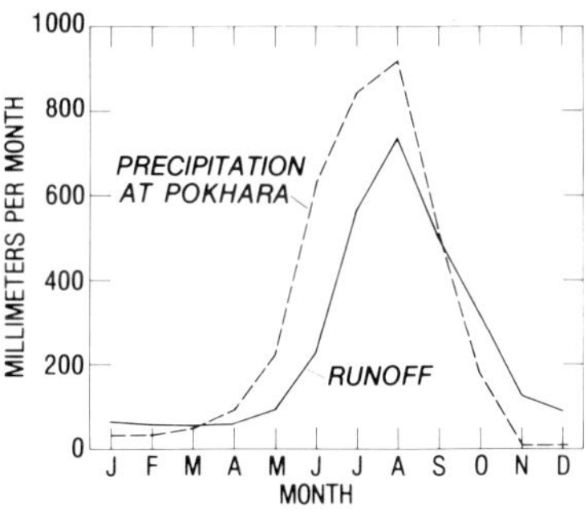

Fig. 2.3. Average monthly precipitation and runoff, Seti Khola, Nepal.

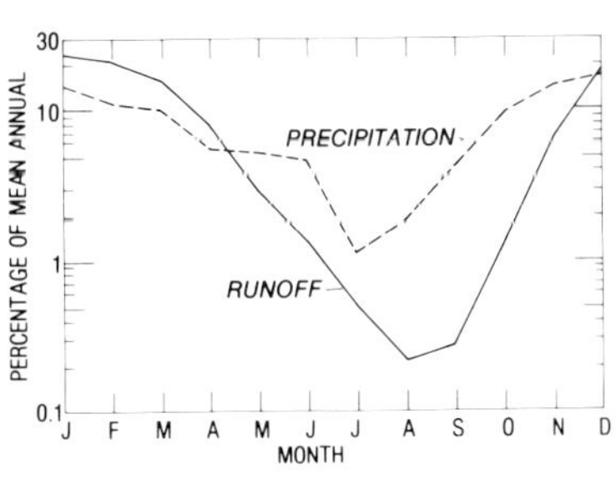

Fig. 2.4. Monthly distribution of precipitation and runoff, Tualatin River, Oregon.

effect of inputs at various times and the lag in runoff from headwaters is a reduction in the variability of downstream flow among seasons. This is illustrated by the seasonal distribution of flow in the Amazon River and of the precipitation at two sites in the basin (Fig. 2.5).

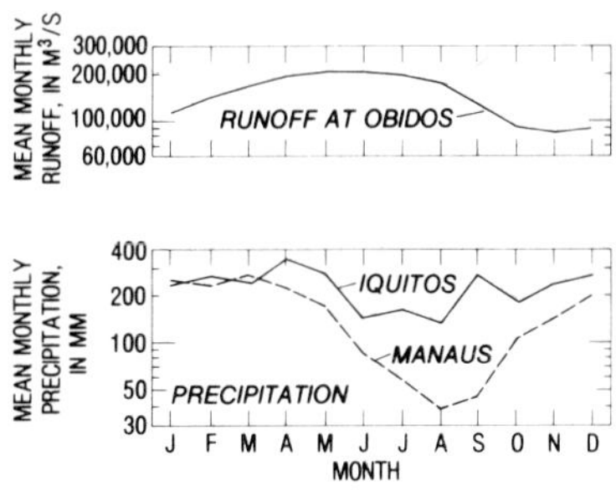

Fig. 2.5. Seasonal distribution of runoff of Amazon River and of precipitation at two sites in the basin.

The seasonal distribution of precipitation also influences the low-flow characteristics of a stream. Frequent storms throughout the year, and especially during the late summer, tend to maintain streamflow whereas, in a climate characterized by little or no rain during summer, the late summer streamflows may be small or zero. The ephemeral stream, one which flows only in direct response to precipitation, and the intermittent stream, which flows for only part of each year on the average, are typical of regions with dry summers.

The above comparisons of seasonal runoff to precipitation have been for average conditions. Variability of weather from year to year accounts for the considerable variation in monthly runoff, in flood peaks, and in low flows. The annual low flow may result from no precipitation for a considerable period or from small amounts of precipitation on each of many days of an extended period, no amounts large enough to do more than supplement the soil moisture. A long period of no precipitation causes a drought both hydrologically and agriculturally, whereas frequent small rainfalls may not produce runoff yet may maintain adequate soil moisture for plants.

Annual precipitation on a river basin varies from year to year. The resulting runoff is more variable than the precipitation because the percentage of precipitation that becomes runoff decreases with decreasing precipitation. Figure 2.6 shows a long record of annual runoffs for Red River of the North, North Dakota, USA. The period of low runoff in the 1930's resulted from several consecutive years of below normal precipitation. More typical are the annual mean runoffs shown in Figure 2.7. The large year-to-year variability in runoff of Rio Guadalquivir in southern Spain is the result of the highly variable

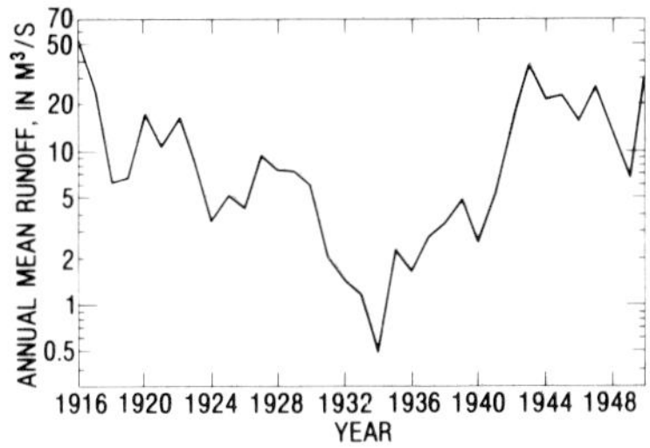

Fig. 2.6. Annual runoff of Red River of the North, North Dakota.

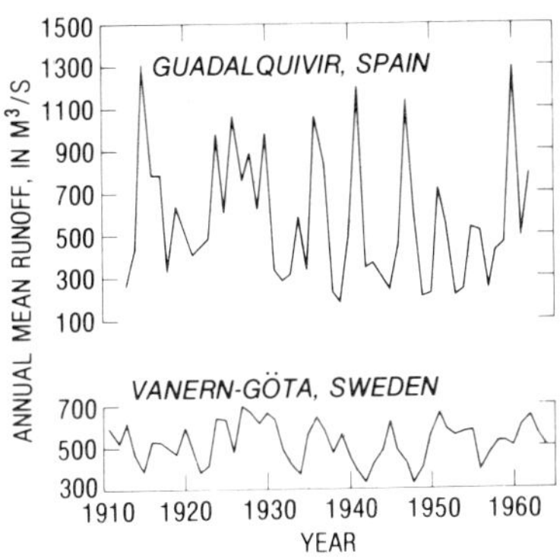

Fig. 2.7. Variability of annual runoffs in a humid region (Sweden) and in a semiarid region (Spain).

annual precipitation characteristic of a semiarid region. In contrast, the annual runoffs of Vanern-Gota, Sweden, in a cool humid climate, are much less variable.

2.2.2 How temperature modifies runoff

Temperature influences runoff in two principal ways: (1) high temperatures increase evapotranspiration and thus reduce the amount of runoff from precipitation, and (2) freezing and thawing change the timing of runoff by delaying the response of runoff to precipitation and by modifying the flow of water already in the channel.

The amount of moisture returned to the atmosphere by evaporation and by transpiration of plants depends on climate, principally temperature, sunshine, and wind, and on the supply of moisture. Frequent rains provide a nearly continuous supply of moisture, but infrequent storms leave little moisture for evapotranspiration for long periods. Thus both potential evapotranspiration and the difference between potential and actual evapotranspiration depend on the climate.

Annual evapotranspiration, or annual water loss, can be quantified as the difference between precipitation on a basin and the resulting runoff. Figure 2.8 shows water losses plotted against mean annual temperature for basins with

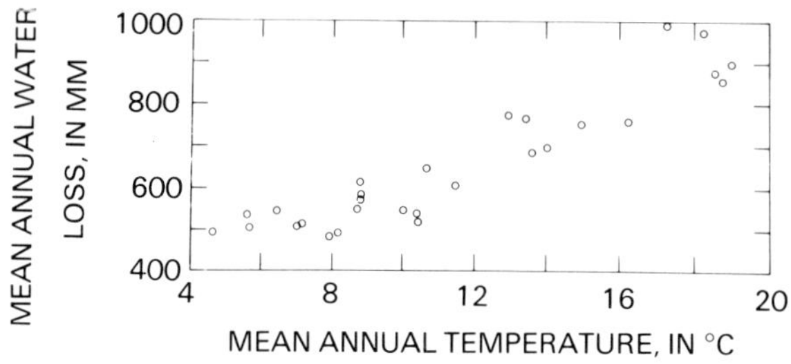

Fig. 2.8. Annual water loss as a function of temperature, eastern USA (from Williams, 1940).

mean annual precipitation greater than 500 mm. Annual water loss is a moderate percentage of the annual precipitation in a humid region but a very large percentage in regions of low precipitation. Thus the annual runoff is highly variable where precipitation is low (Red River of the North and Rio Guadalquivir in Figs. 2.6 and 2.7). In temperate and tropical arid regions, potential evapotranspiration is larger than precipitation except for short periods during and following rainstorms. Thus there are no perennial streams in such regions.

Temperature and thus evapotranspiration rates change seasonally in most parts of the world. Figure 2.9 shows that losses during that summer are larger than

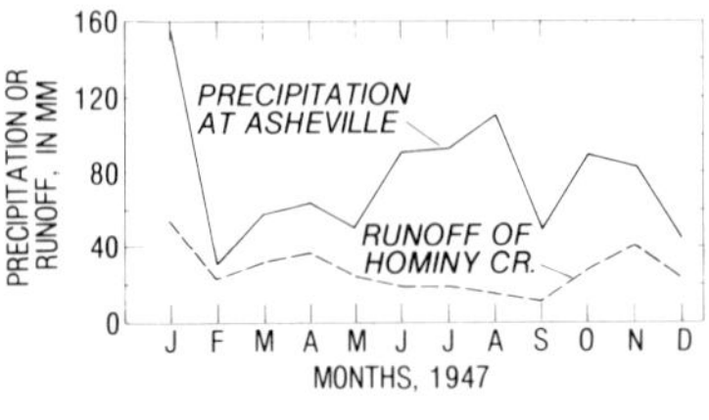

Fig 2.9. Seasonal relation of precipitation and runoff showing the large evapotranspiration losses in summer in North Carolina. (January precipitation probably was snow.)

at other times of the year. The precipitation on the Hominy Creek basin probably is not the same as at Asheville but the monthly trends should be similar. The difference between precipitation and runoff for a single month is only a rough approximation of loss because of the carryover of water from one month to the next.

Diurnal variations in evapotranspiration often produce diurnal fluctuations in the flow of small streams where the temperature difference between night and a sunny day is large. Figure 2.10 shows the discharge hydrograph during such a

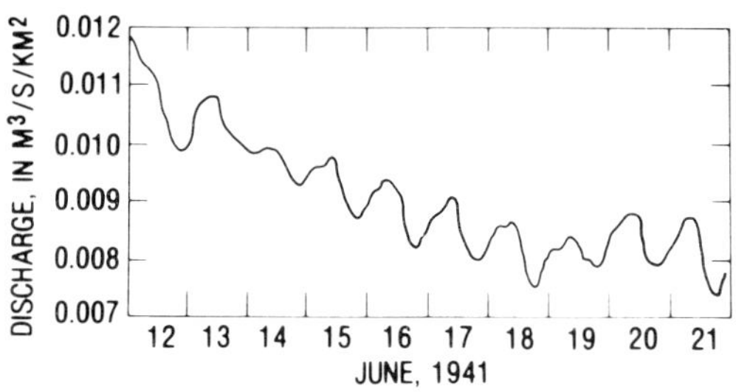

Fig. 2.10. Diurnal fluctuations in streamflow caused by variable evapotranspiration losses (modified from Dunford and Fletcher, 1947).

period. The effect of evapotranspiration on streamflow is sometimes manifested in the drying up of a small stream during summer and its subsequent resumption of flow in the absence of rain as temperature decreases in the fall months. The magnitude of evapotranspiration may also be indicated by a sharp increase in streamflow after the first hard frost of autumn kills the riparian vegetation.

The within-year distribution of runoff of Arctic streams is determined by temperature, not precipitation, because the winter precipitation remains in storage as ice or snow until summer. The mean monthly runoffs of a typical Arctic stream, Yana River, Dzanhky, USSR, are shown in Figure 2.11 along with the precipitation and temperature at Verhoyansk. Low flows during winter result from freezing of nearly all water in the channel; the base flow of a stream draining a permafrost regime is negligible.

A less extreme range in monthly flows of a northern stream occurs on Yukon River, Canada and Alaska, where the flow under ice cover recedes throughout the winter but is never less than several hundreds of cubic meters per second. Temperature is again the major factor in seasonal distribution of flow although the maximum precipitation does coincide with the summer high flow.

A sharp rise in temperature following a prolonged cold spell will release the heavy ice cover on a river. The resulting ice jams may damage structures in and along the river and cause overbank flooding. Northward flowing streams are most

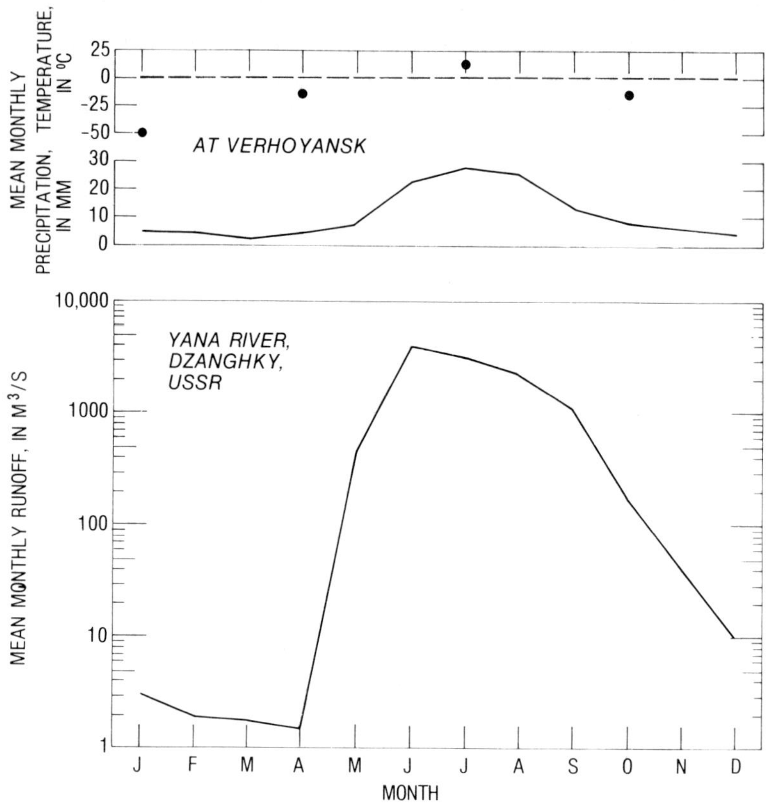

Fig. 2.11. Seasonal distribution of runoff of an arctic stream showing the effect of temperature.

susceptible to ice jams because the southern headwaters thaw before the northerly downstream reaches of the channel.

Small streams in northern climes freeze to the bottom during the winter. In certain nonpermafrost reaches, ground water continues to feed the stream but because this water cannot flow down the stream channel it flows over the ice until it freezes. Thus, the ice cover on the stream and overbank is built up throughout the winter. This icing may encompass roads, block culverts, and invade buildings along a stream.

Temperature effects on streamflow where complete ice cover does not occur throughout the winter are somewhat different than those above. The effect on a large stream is shown in Figure 2.12 (Simons, 1953). Extremely cold weather for a few days puts water in storage as ice in the channel and reduces the flow.

Recovery of usual flow follows an increase in temperature. The big increase in flow in early February was caused by above-freezing temperatures and rain.

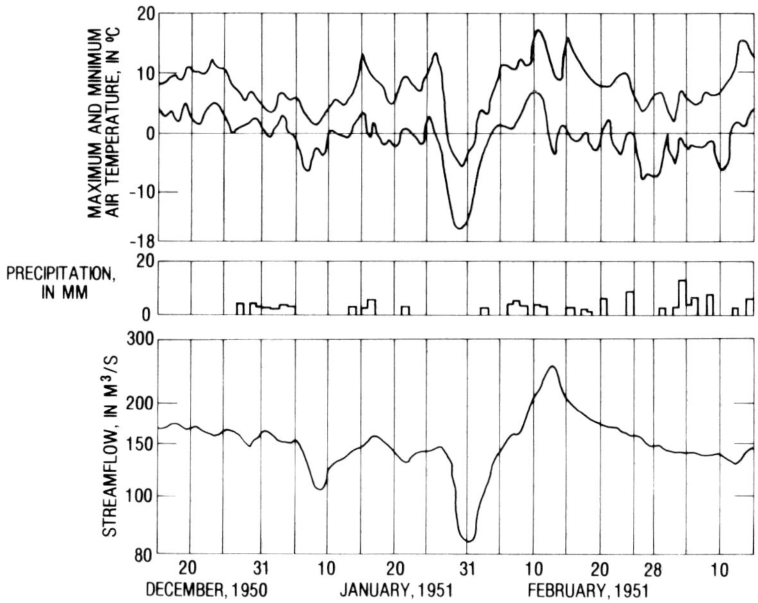

Fig. 2.12. Effect of temperature on streamflow, Salmon River, Idaho.

In more southerly climes, and generally at high elevations, the temperature during winter may range from much below freezing at night to above freezing during the day. During clear, cold nights anchor ice forms on the rocks of the streambed because of loss of heat by radiation to the atmosphere. Radiation from the sun the next day warms the rocks and releases the ice. Anchor ice does not form under surface ice cover. The formation and subsequent release of anchor ice causes substantial fluctuations in water level in the stream (Fig. 2.13) and smaller fluctuations in discharge.

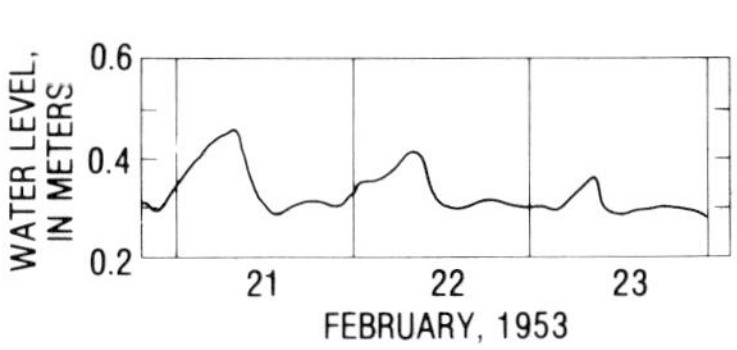

Fig. 2.13. Water level fluctuations in a small stream due to formation and release of anchor ice (modified from Moore, 1957).

Glaciers and snowfields are common in mountainous regions at high latitudes. Runoff from these features is more closely related to temperature than to the previous winter precipitation. Streams with glaciers in their headwaters have typically high summer runoff. Hydrographs of a glacial and a nonglacial stream in Washington State, USA, in Figure 2.14, show the opposite effects of

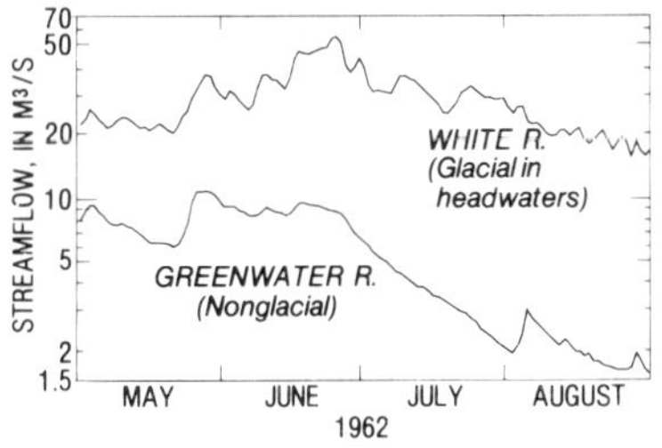

Fig. 2.14. Hydrographs showing the opposite effects of a dry July on a glacial and a nonglacial stream.

temperature on runoff during a typically dry July. High temperature increases the losses from both basins, but increases glacier melt in White River basin. During other seasons the flows of the two streams respond similarly to temperature and precipitation.

Glaciers sometimes cause unusual floods by impounding water and then releasing it abruptly. From at least 1948 to 1962, Knik Glacier, Alaska, advanced across Knik River each winter and temporarily stored the spring runoff. Destruction of the glacier dam each summer by melting and erosion released the stored water causing an outstanding flood on Knik River. Figure 2.15 shows hydrographs for years with and without the glacier effect. Annual runoffs for these 2 years are virtually the same.

In temperate regions precipitation may occur as snow during the winter and be released in spring or early summer, particularly if the streams head at high elevations. Maximum streamflows often occur as a result of snowmelt. The amount of runoff from a snowpack depends on the weather during the melt period. A rapid melt will produce the most runoff. An extended melt period caused by cool temperatures permits more loss by ablation and by infiltration.

2.2.3 Climatic differences along a stream

Some streams drain areas having one climatic regime in the headwaters and another in the downstream part of the basin. Commonly the principal runoff is generated in the headwaters which has the higher precipitation; then the downstream runoff characteristics are a composite.

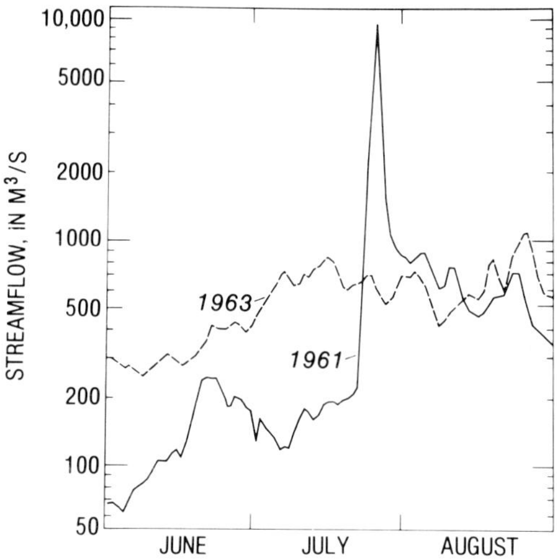

Fig. 2.15. Hydrographs for Knik River, Alaska, showing the effect of impoundment and release of water by a glacier (1961), and a year with no impoundment (1963).

Two flood periods per year are produced on Merced River, California, one from low-elevation rainfall in winter and the second from the melting of the snowpack in the Sierra Nevada in late spring. Figure 2.16 shows the annual flood-frequency curves for the two types of floods (Crippen, 1978). The snowmelt flood is the larger in most years but the major floods are produced by rainfall.

Commonly, a stream that heads in the high mountains will reach its annual maximum discharge in its upstream reaches from snowmelt while further downstream the annual maximum discharge may be either from snowmelt or from rainfall or, occasionally, a combination of the two. The annual runoff at a downstream site on such a stream may be larger or smaller than the runoff at some point upstream depending on the downstream climate.

Typically, but not always, the headwaters of a stream receive more precipitation than the lower reaches. If the disparity in precipitation between the upper and lower basins is not too great, or if the lower basin is small relative to the upper basin, then runoff will increase downstream. If there is little or no contribution from the lower basin, two results can occur: (1) a large upstream runoff will be transmitted through the downstream basin with only relatively moderate losses, or (2) upstream runoff may be greatly (or entirely) dissipated by evapotranspiration and by infiltration through the channel bed in the lower basin. The Nile River is an example of the first. Many examples of the second are known in arid regions although the flow is usually reduced by manmade diversions in addition to the natural losses. Many streams that flow

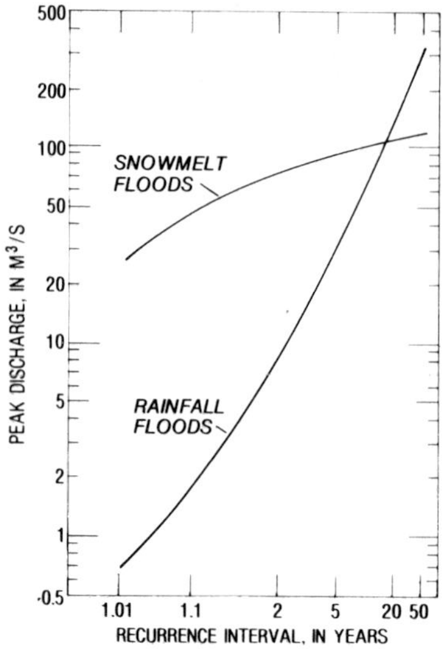

Fig. 2.16. Flood-frequency curves of snowmelt and rainfall floods, Merced River, California.

into arid regions terminate in closed lakes; Great Salt Lake, Dead Sea, Lake Chad, and Caspian Sea are well known examples. Other streams terminate in playas which may be dry for long periods. Still other streams, such as the many wadis draining toward the interior of the southern Arabian Peninsula, gradually disappear into the desert.

2.3 EFFECTS OF GEOLOGY

Under the same climatic influences, greatly different streamflow regimes are produced from drainage basins having different soils and rocks. The soil or other surficial material determines how much and at what rate the precipitation will infiltrate and thus what proportion will become overland runoff. Some of the infiltrated water will be evaporated and the rest will move through the rocks either to a stream channel or, in some arid regions, to a zone of saturation below any stream channel. The character of the rocks determines how the water moves underground and at what rates.

Rapid surface runoff and little base flow are associated with low infiltration rates. Conversely, the streamflow is much less variable from a basin with a high infiltration rate. The following examples demonstrate how geology modifies streamflow; in most basins geology may be only one of several significant basin characteristics and its effect on streamflow may not be readily apparent.

Regions in which practically all the precipitation soaks into the ground usually are drained by streams with little variation in flow. The Sand Hills of Nebraska is such a region. Figure 2.17 compares the monthly mean flows of

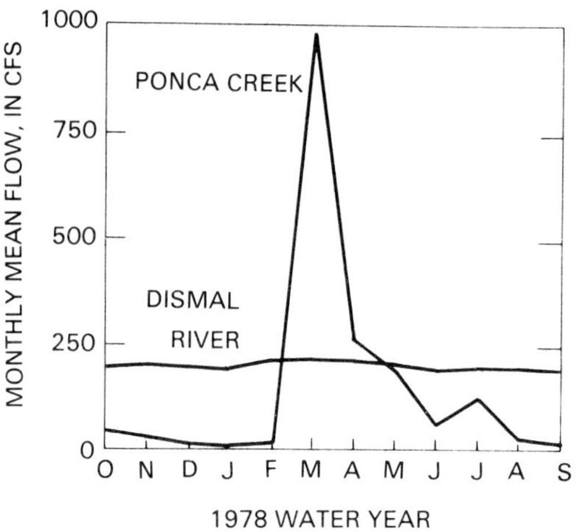

Fig. 2.17. Monthly mean flows of two Nebraska streams. Dismal River is in the Sandhills Region.

Dismal River, a Sand Hills stream, with Ponca Creek, which is not. A less extreme difference is shown between two North Carolina streams, Drowning Creek which drains a sandy area, and Uwharrie River whose basin is less permeable (Fig. 2.18). Variation in daily flows of two geologically-different streams is shown in Figure 6.3 (Chapter 6).

The Pahsimeroi River in Idaho flows through an alluvial valley between high mountains. Snowmelt runoff from the tributaries sinks into the ground before reaching the river. Low-elevation springs in this basin usually reach their maximum discharges in August or September. The moderating effect of the alluvial fill is so great that many of the annual flood peaks on Pahsimeroi River occur in November or December although all other gaged streams in the vicinity peak in spring or early summer.

The Yucatan Peninsula of Mexico is a flat, limestone plateau in which there are few stream channels. Practically all of the precipitation infiltrates into the ground and then drains directly to the ocean. Some sandy coastal areas in humid parts of the world also are without stream channels.

Basins with similar infiltration characteristics may have very different outflow regimes due to differences in the rate of movement through the ground and in the amount of storage in the aquifer. Troxell (1953) showed that a major

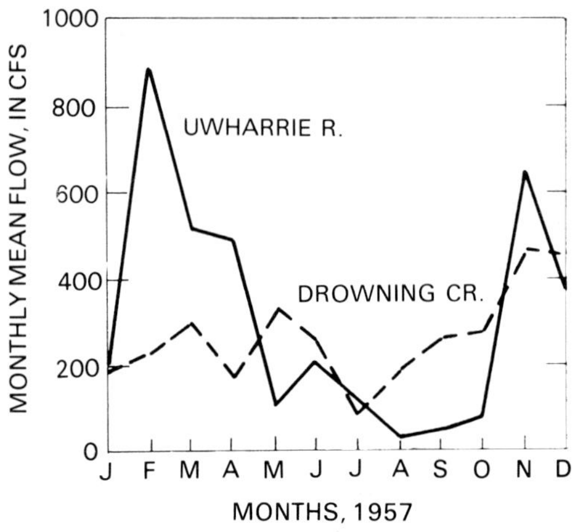

Fig. 2.18. Monthly mean flows of two North Carolina streams showing the difference due to basin infiltration rates.

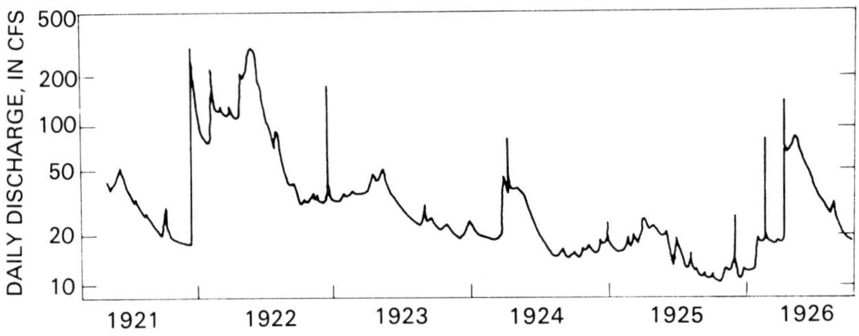

Fig. 2.19. Hydrograph for Mill Creek, California, showing that the major recharge in 1922 contributed to the base flow for the next three years (after Troxell, 1953).

recharge of the small, steep basin of Mill Creek in southern California contributed to the base flow for several succeeding years (Figure 2.19), an indication of a large ground water body that drained relatively slowly to the stream. McDonald and Langbein (1948) concluded that the ground water storage in the basin of Metolius River, Oregon was about three times the annual runoff. Metolius River drains a basalt basin and its flow is uncommonly steady. On the other hand the flow of Drowning Creek (Figure 2.18) responds rather promptly to precipitation, indicating a more limited aquifer capacity. Meinzer (1949) describes the occurrence, origin, and discharge of ground water and gives many examples.

Some basins transmit infiltrated water to stream channels very rapidly through solution channels, porous lava, or coarse alluvial material. The hydrograph of Jack Daniel Spring in Figure 6.2 (Chapter 6) responds rapidly to precipitation presumably because the water is transmitted through openings in the limestone. Related occurrences are the disappearance and subsequent reappearance of stream channels in limestone or volcanic terranes. Figure 2.20

Fig. 2.20. Popo Agie River, Wyoming, where it disappears into the ground.

shows the Popo Agie River in Wyoming going underground from which it emerges within a short distance. Rogue River in Oregon disappears in lava tubes and then reappears. And Lost River, in the headwaters of the Potomac River, goes underground near Wardensville, W. Va. in a limestone terrane.

Streams flowing in porous material may lose water through the channel bed and banks in certain reaches. Thus the surface flow at a particular point on the channel may be only a part of the total flow past that section. Gaging stations are usually located above outcrops which force the total flow to the surface. Furthermore, discharges of ground water to a stream often are not distributed uniformly along the channel. Large increases or decreases in ground-water contributions may occur within a short distance along a reach as indicated by the flow profile of a short reach of Spokane River (Figure 2.21). There are no appreciable surface water contributions in this reach.

The contribution of ground water to a stream often increases with distance downstream as the stream cuts deeper into the aquifer. This effect has been

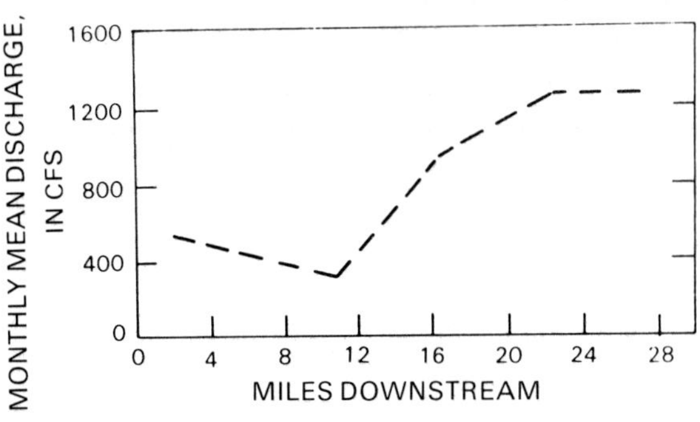

Fig. 2.21. Effect of ground water on the flow of Spokane River from Post Falls, Idaho, to mouth of Latah Creek in Spokane, Washington, September 1950.

observed on some Long Island, New York, streams whose base flows increase considerably as they approach the ocean. For example see Figure 2.22.

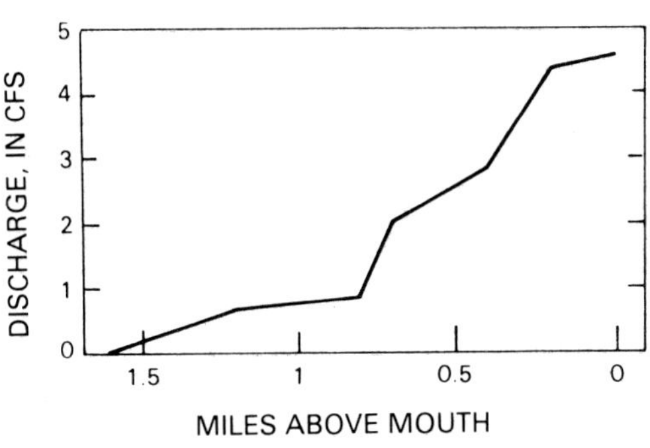

Fig. 2.22. Base flow of Carman Creek, Long Island, New York, showing increase toward mouth, 12/28/78.

Basin geology may permit substantial amounts of ground water to move across topographic boundaries, causing an unequal distribution of runoff with respect to surface-water drainage areas. The large difference in yield per square mile of two small streams in a limestone terrane in southeastern Idaho (Fig. 2.23) is attributed to this cause. Drainage areas of Cub Creek and Bloomington Creek are are 19.4 and 22.1 square miles respectively. Bloomington Creek has large springs in its headwaters.

In arid and semiarid regions where the water table is below stream channels the permeability of the stream bed determines the rate at which streamflow seeps into the ground. Transfer of water between the surface and the ground also may

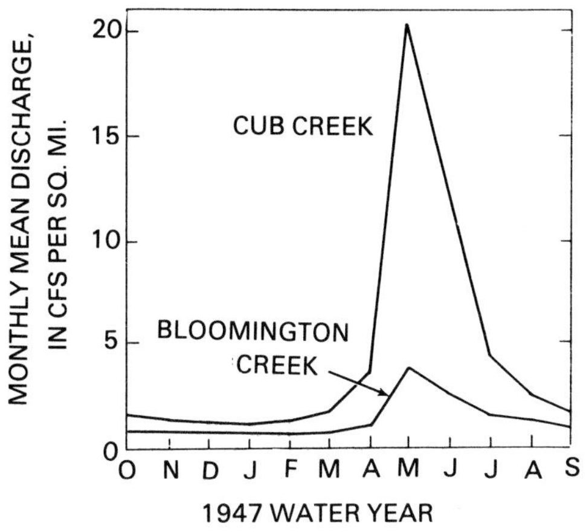

Fig. 2.23. Difference in yield from two adjacent Idaho drainage basins is attributed to movement of ground water across topographic divides.

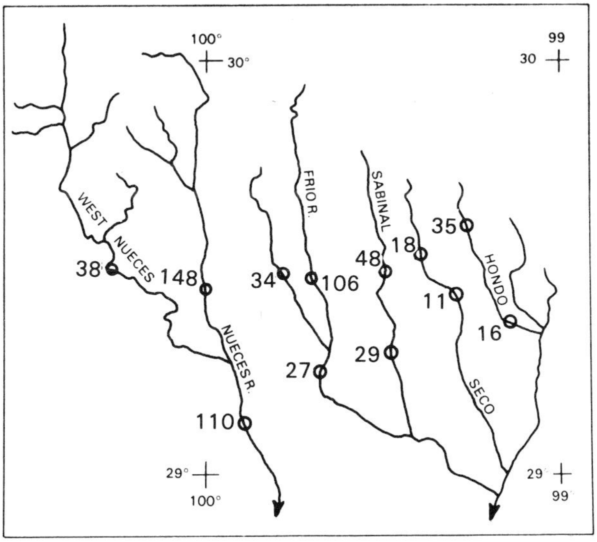

Fig. 2.24. Reduction in mean flows due to streams crossing the Balcones Fault area, Texas.

occur at geologic faults. Figure 2.24 shows mean flows at gaged sites in the upper basins of the Nueces and Frio Rivers, Texas. The flow losses, indicated by the mean flows, occur where the streams cross the Balcones Fault area.

2.4 EFFECTS OF TOPOGRAPHY

In mountainous basins, land slopes are steep, stream gradients are high, and precipitation reaches the channels quickly. Many studies have shown that flood peaks increase with channel slope. Precipitation reaches the channels more slowly where basin slopes are gentle; consequently streams take longer to rise to a peak and the peaks tend to be flattened. A flood peak may move downstream two or three times as fast on a high-gradient stream as on one of low gradient. And low gradient channels tend to be in wide valleys where large riparian areas are inundated by significant floods; the resulting channel storage reduces the peak discharge although much of the overflow is returned to the stream rather promptly. Overflows of some channels in very flat topography may never return to the main channel. This natural diversion reduces both the flood peak and the yield downstream.

Lakes, swamps, and marshes are common in basins of low relief. The natural storage in these areas generally reduces the downstream flood peaks. Evaporation from the water surfaces and evapotranspiration from the perimeters of the wetland areas and from the shallow ground water reduce basin runoff substantially at times. The variability in storage due to weather patterns results in a high variability of annual low flows downstream. For example the annual low flows of Suwannee River, which drains the 650 square-mile Okefenokee Swamp in Georgia, range from about 700 cfs to zero at the Georgia-Florida State line.

In extremely flat regions where the drainage divides are not well defined water may flow in one or the other of two directions. Usually a shallow pond or wetland is the headwater of both streams. An example is the headwaters of Curlew Creek and San Poil River in eastern Washington State.

Channels in flat, coastal areas tend to be interconnected and their flows depend on tidal action and on the locations of storms and inflows. The flow in a particular reach may be in one direction at one time, and in the opposite at another; between those times the channel may be full of water that is not moving.

Stream pattern can be considered a topographic feature. It affects the flood characteristics along the main channel. Tributaries entering a channel at regular intervals of distance will produce a flood that increases downstream if the tributary peaks are substantial and more or less concurrent with the peak on the main channel. Just below the confluence of two streams of similar size and regime, the combined flood peak may approximate the sum of the two contributing peaks unless the two peaks arrive at different times, in which case a double peak may result. In a long channel reach without tributaries, or with tributaries that do not contribute at the same time as the headwaters, a flood generated upstream will be attenuated; see Figure 8.12 (Chapter 8).

A common indicator of stream pattern is some index of basin shape but shape is often a poor descriptor of flood potential. Unless the pattern is extreme, and the basin hydrology is more or less homogeneous, the effect of pattern probably is small and will be masked by other basin characteristics.

The effect of stream pattern on mean flow or low flows is either negligible or to small to detect.

A topographic feature that affects streamflow in arid regions is the closed basin from which no surface flow leaves. Water generated in a closed basin flows into a terminal lake or a sink. Well known large closed basins are the Great Basin in western United States, the Jordan River basin in the Arabian Peninsula, and the Caspian Lake Basin in Asia. Considerable streamflow is generated in these basins but none of it reaches the oceans.

Small closed basins are common in semiarid regions of the world. The contributing drainage areas of some streams in western United States are less than the areas enclosed by the topographic boundaries because of closed basins within those boundaries. Some of these small closed basins may overflow during extreme floods. Basins are closed because of low precipitation; if the precipitation were high enough the basin would become a lake and drain to an ocean.

REFERENCES

Crippen, J.R., 1978, Composite log-type III frequency-magnitude curve of annual floods: U.S. Geol. Survey Open- File Report 78-352, 5 p.

Crippen, J.R., and Bue, C.D., 1977, Maximum floodflows in the conterminous United States: U.S. Geol. Survey Water-Supply Paper 1887, 52 p.

Dunford, E.G., and Fletcher, P.W., 1947, Effect of removal of streambank vegetation upon water yield: Am. Geophysical Union, Trans., Vol. 28, No. 1, Feb 1947, p. 105-110.

L'vovich, M.I., 1980, World water resources and their future (translation edited by R.L. Nace): Am. Geophysical Union, Washington, D.C., 416 p.

McDonald, C.C. and Langbein, W.B., 1948, Trends in runoff in the Northwest: Trans. Am. Geophysical Union, Vol. 29, No. 3, June 1948, p. 387-397.

Meinzer, O.E., 1949, Hydrology: New York, Dover Publications, 712 p.

Moore, A.M., 1957, Measuring streamflow under ice conditions: Am. Soc. Civil Engineers Proc., Jour. Hydraulics Div., v. 83, HY1, Paper 1162, 12 p.

Riggs, H.C., 1980, Climatic factors of runoff, in Pollution and Water Resources: Columbia University Seminar Series, V. XIII, Part III, Pergamon Press, p. 73-83.

Simons, W.D., 1953, Concept and characteristics of base flow in the Columbia River basin: Western Snow Conf. Proc., 21st meeting, Boise, Idaho, p. 57-61.

Troxell, H.C., 1953, The influence of ground-water storage on the runoff in the San Bernardino and eastern San Gabriel Mountains of southern California: Trans. Am. Geophysical Union, Vol. 34, No. 4, p. 552-562.

Williams, G.R., 1940, Natural water loss in selected drainage basins: U.S. Geol. Survey Water-Supply Paper 846, 52 p.

Chapter 3

COLLECTION OF HYDROLOGIC DATA

3.1 STREAMFLOW

Continuous streamflow records are obtained at a gaging station at which the stream stage (water-surface height above some datum) is either read frequently or is recorded continuously. Discharge is measured, usually by current meter, at various stages for defining the stage-discharge relation (rating curve) which is used to convert the stage record to a discharge record. Only the general procedures are given here; for more detail see WMO (1980) or Rantz (1982).

3.1.1 Stage measurement

Gaging stations are of several types. The simplest consists of a series of staff gages (Fig. 3.1) on which an observer reads the stage one or more times a

Fig. 3.1. Staff gages on Rio Cia, Brazil.

day. Recording gages either sense the water surface in a stilling well hydraulically connected to the stream or by means of a gas-purge (bubble gage) system which measures the pressure on an orofice permanently mounted in the stream. Recording gaging stations also have staff gages (Fig. 3.2) for verifying the stages being sensed by the equipment. A simplified stilling-well installation is shown in Figure 3.3 and a gaging station structure in Figure 3.4.

A bubble-gage installation consists of an instrument shelter on a concrete slab on the river bank. A tube connects the instrument to the orofice in the stream. Nitrogen gas is bubbled through the orofice to actuate the sensor.

Fig. 3.2. Low-water staff gage.

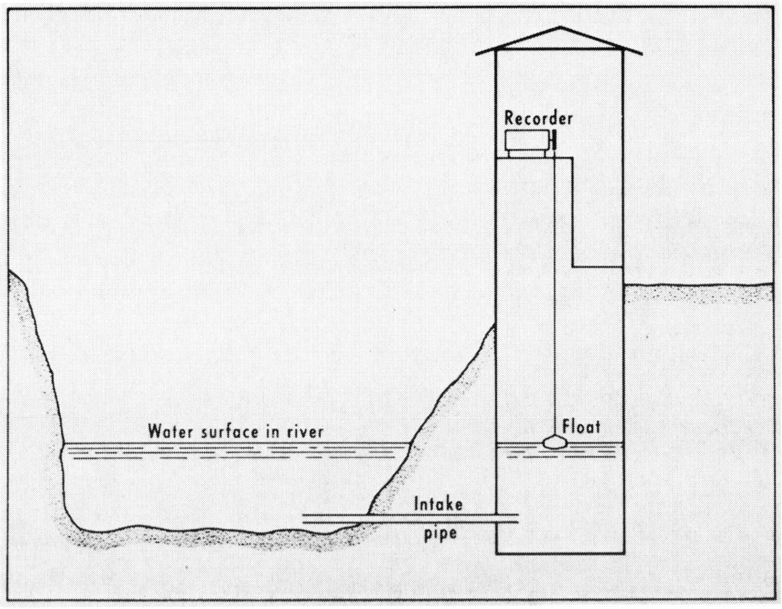

Fig. 3.3. Simplified stilling-well installation.

Recording instruments are either analog or digital. The latter produces a punched-tape record for processing by digital computer; the punch interval commonly used is 15 minutes but this can be as short as 5 minutes or as long as an hour. The two types of recorders are shown in Figures 3.5 and 3.6.

A gaging station should be located above a stable section of channel in order that the relation between stage and discharge be well defined and unchanging

Fig. 3.4. Gaging station.

Fig. 3.5. Analog water-stage recorder.

with time. The feature of the channel that maintains a more-or-less stable stage-discharge relation is called the control. The control may be at a section, such as a stable riffle (Fig. 3.7) or it may be a fairly long reach of the

Fig. 3.6. Digital water-stage recorder.

Fig. 3.7. Natural section control.

channel itself. Artificial section controls such as low dams are sometimes constructed where natural channel features are not suitable. Artificial controls are expensive, especially for large streams, and are hard to maintain in erodible channels carrying heavy sediment loads.

3.1.2 Discharge measurement

Discharge measurements of streams are usually made by current-meter. The procedure consists of (1) measuring the width, depth, and velocity of flow in each of several subsections of a stream cross section, (2) computing the discharge in each subsection as the product of area and mean velocity, and (3) summing the partial discharges to obtain the total. Referring to Figure 3.8, the depth at each of the selected verticals is measured by sounding and the width of each subsection is computed from the spacing of the verticals. At each vertical the mean velocity is obtained from one or more velocity observations by

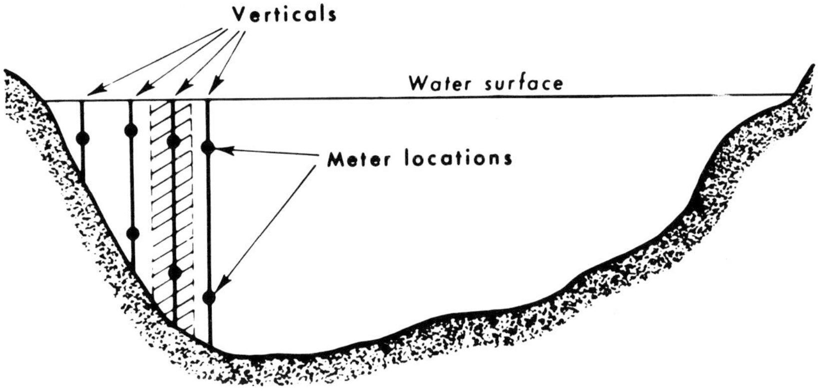

Fig. 3.8. Stream cross section showing meter locations for a discharge measurement.

current meter. Many studies have demonstrated that the mean of the velocities at 0.2 and 0.8 of the depth from the water surface is virtually the mean velocity in the vertical. Likewise the velocity at 0.6 depth very nearly equals the mean in the vertical. Velocity observations are usually made at 0.2 and 0.8 of the depth in each vertical where the depth is adequate.

The basic equipment needed for a discharge measurement consists of a current meter, a device for indicating the revolutions of the meter, a stop watch, and some means of measuring depth and width and of holding the meter in the proper locations for velocity observations. In shallow streams, measurements are made by wading; the current meter is mounted on a wading rod which is used to measure depth and to position the meter in the vertical. Horizontal control is maintained by a tape or beaded wire stretched across the stream. See Figure 3.9. In use, the number of revolutions of the meter rotor is obtained by an electrical circuit which produces clicks in an earphone or registers on a counting device. Elapsed time is measured by a stopwatch. These data are translated to velocity through the meter rating table. Figure 3.10 shows notes of a discharge measurement.

Fig. 3.9. Wading equipment and a measurement in progress.

Deep streams are measured from a bridge, cableway, or boat. The meter is suspended on a cable above a sounding weight which is used for depth measurement and to hold the current meter at the correct position in the vertical for velocity observation. Sounding weights used range from 15 to 300 pounds or more depending on the depth and velocity of the stream. Methods of gaging are shown in Figures 3.11, 3.12, and 3.13. See Buchanan and Somers (1969) for a more complete description of measurement techniques including measurement under ice cover.

The moving-boat method is similar to the above method in that it uses the velocity-area approach to determining discharge but it differs in the method of data collection; the depths and velocities at each observation point are obtained while the boat is rapidly traversing the cross section. Thus a measurement of discharge by this method can be made on wide streams in a few minutes at sites without fixed facilities. However, in practice 6 traverses usually are made and the results averaged. The equipment (Fig. 3.14) consists of a sonic sounder, a vane with angle indicator, a component propellor-type current meter, and a maneuverable small boat.

The traverse of the cross section is made without stopping and data are collected at regular intervals. The boat operator maintains course by "crabbing" into the direction of flow sufficiently to keep on line. The force exerted on the current meter is the combination of two forces acting simultaneously, one force due to the movement of the boat and the other from the streamflow normal to the path. Additional information needed is the vertical

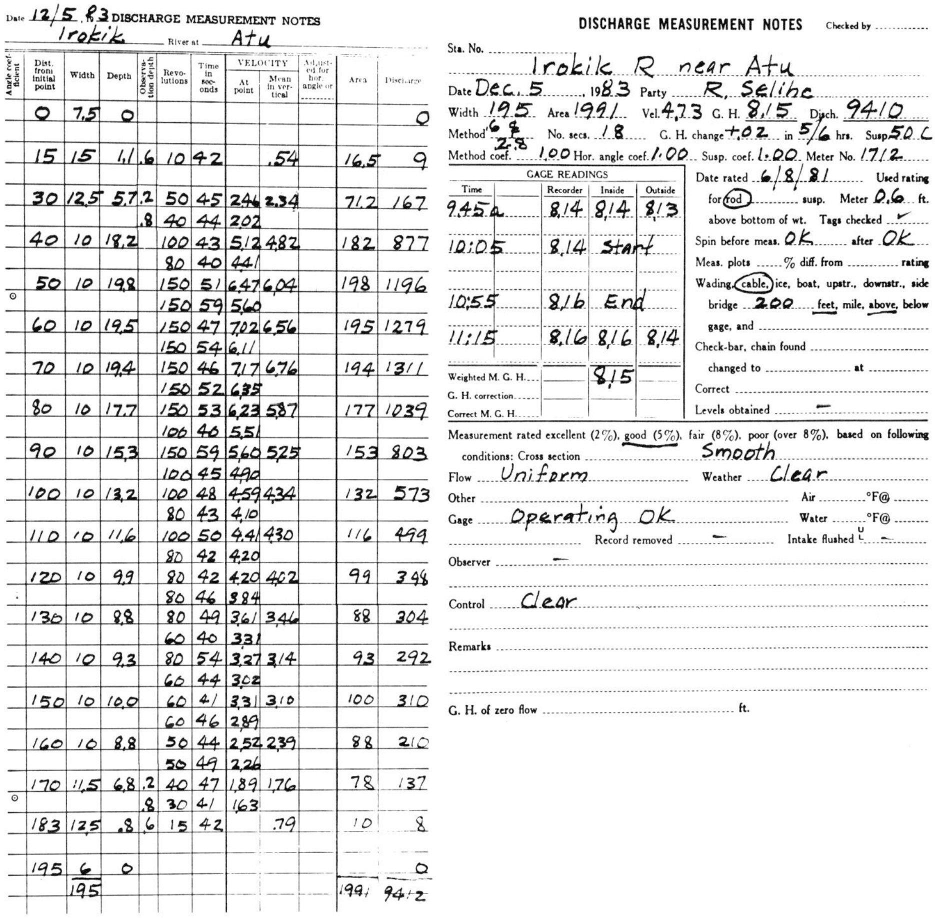

Date 12/5 83 DISCHARGE MEASUREMENT NOTES

Irokik River at Atu

Angle coefficient	Dist. from initial point	Width	Depth	Observation depth	Revolutions	Time in seconds	Velocity At point	Velocity Mean in vertical	Adjusted for hor. angle or	Area	Discharge
	0	7.5	0								0
	15	15	1.1	.6	10	42		.54		16.5	9
	30	12.5	5.7	.2	50	45	2.46	2.34		71.2	167
				.8	40	44	2.02				
	40	10	18.2		100	43	5.12	4.82		182	877
					80	40	4.41				
	50	10	19.8		150	51	6.47	6.04		198	1196
⊙					150	59	5.60				
	60	10	19.5		150	47	7.02	6.56		195	1279
					150	54	6.11				
	70	10	19.4		150	46	7.17	6.76		194	1311
					150	52	6.35				
	80	10	17.7		150	53	6.23	5.87		177	1039
					100	46	5.51				
	90	10	15.3		150	59	5.60	5.25		153	803
					100	45	4.90				
	100	10	13.2		100	48	4.59	4.34		132	573
					80	43	4.10				
	110	10	11.6		100	50	4.41	4.30		116	499
					80	42	4.20				
	120	10	9.9		80	42	4.20	4.02		99	398
					80	46	3.84				
	130	10	8.8		80	49	3.61	3.46		88	304
					60	40	3.31				
	140	10	9.3		80	54	3.27	3.14		93	292
					60	44	3.02				
	150	10	10.0		60	41	3.31	3.10		100	310
					60	46	2.89				
	160	10	8.8		50	44	2.52	2.39		88	210
					50	49	2.26				
	170	11.5	6.8	.2	40	47	1.89	1.76		78	137
⊙				.8	30	41	1.63				
	183	12.5	.8	.6	15	42		.79		10	8
	195	6	0								0
		195								1991	9412

DISCHARGE MEASUREMENT NOTES Checked by

Sta. No.

Irokik R near Atu

Date Dec. 5, 1983 Party R. Selihc

Width 195 Area 1991 Vel. 4.73 G. H. 8.15 Disch. 9410

Method 6 & 2.8 No. secs. 18 G. H. change +.02 in 5/6 hrs. Susp 50 C

Method coef. 1.00 Hor. angle coef. 1.00 Susp. coef. 1.00 Meter No. 1712

GAGE READINGS Time		Recorder	Inside	Outside
9.45a		8.14	8.14	8.13
10:05		8.14	Start	
10:55		8.16	End	
11:15		8.16	8.16	8.14
Weighted M. G. H.			8.15	
G. H. correction				
Correct M. G. H.				

Date rated 6/8/81 Used rating for (rod) susp. Meter 0.6 ft. above bottom of wt. Tags checked ✓

Spin before meas. OK after OK

Meas. plots % diff. from rating

Wading, (cable), ice, boat, upstr., downstr., side bridge 200 feet, mile, above, below gage, and

Check-bar, chain found changed to at

Correct

Levels obtained —

Measurement rated excellent (2%), good (5%), fair (8%), poor (over 8%), based on following conditions: Cross section Smooth

Flow Uniform Weather Clear

Other Air °F@

Gage Operating OK Water °F@

Record removed — Intake flushed —

Observer —

Control Clear

Remarks

G. H. of zero flow ft.

Fig. 3.10. Discharge measurement notes.

velocity curve for the section and the measured width of the section. Details of the method are given by Smoot and Novak (1969). The method is suitable only on wide rivers and has been used to measure many of the major rivers of the world.

3.1.3 Rating Curve

When discharge measurements are available throughout the ordinary range of stage, the relation is defined graphically, usually on log paper. If the gaging-station control is stable, and measuring conditions are good, the rating curve will be fairly well defined as is the one in Figure 3.15. More commonly

Fig. 3.11. Gaging from a bridge (during a flood at left).

Fig. 3.12. Measuring from a cableway.

the stage-discharge relation will change either gradually or abruptly in response to such factors as aquatic growth, ice formation and release, erosion or deposition by floods, and other natural or man-made changes in the channel. The definition and application of rating curves require an understanding of stream hydraulics and considerable experience. See WMO (1980) or Rantz (1982), and Kennedy (1983).

Fig. 3.13. Jet boat equipped for gaging.

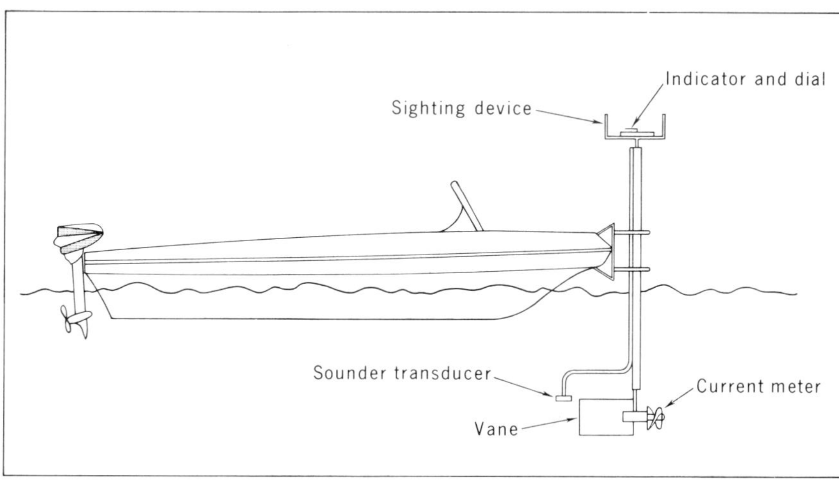

Fig. 3.14. Equipment for measuring by the moving-boat method (From Smoot and Novak, 1969).

A stage-discharge relation can be computed from a survey of the downstream channel and estimates of channel roughness. The technique, known as the step-backwater method, is useful where time does not permit obtaining current-meter discharge measurements throughout the range of stage, or where high accuracy is not required. The method is described in many hydraulic texts and by Davidian (1984) and was verified for this application by Bailey and Ray (1966).

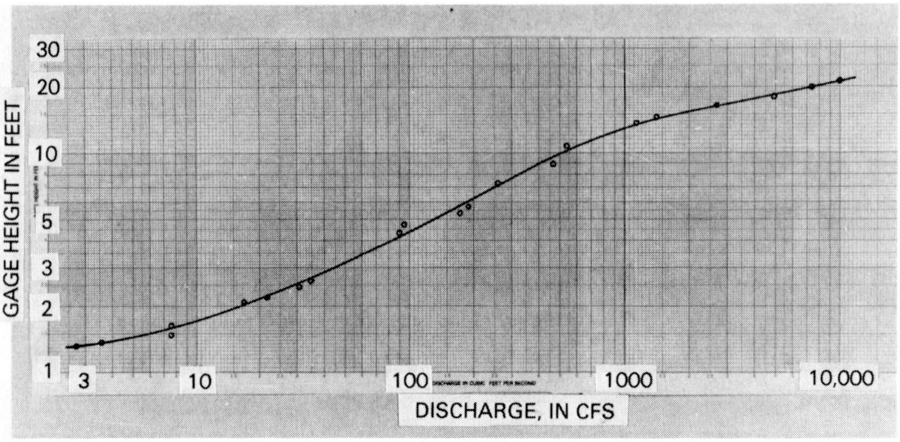

Fig. 3.15 Stage-discharge relation (rating curve).

Streamflow records are sometimes needed on stream reaches affected by variable backwater. The discharge past a section on such a reach is a function of stage and the slope of the water surface through the reach. A continuous stage record at each end of a reach is required to define the slope. The method of analysis depends on the hydraulic conditions. See WMO (1980) or Rantz (1982), and Kennedy (1983) for rating curve theory and details of applications to various channel and flow conditions.

3.1.4 Discharge computation and the hydrograph

Given a continuous stage record or frequent stage observations, and a stage-discharge relation, the discharge can be computed for any particular time within the period of stage record. Daily mean discharges are usually computed although discharges at intervals of an hour or less are used to define the changes during a flood. An annual streamflow record as published by USGS is shown in Figure 3.16.

The daily mean discharges of Figure 3.16 are plotted against time to produce the hydrograph shown in Figure 3.17. Hydrographs show how a stream responds to various events. An uncharacteristic pattern of a segment of a hydrograph may indicate some unnatural flow modification or an error in the record or in the analysis on which the discharge record is based.

Flood hydrographs are usually defined by discharges at short intervals, such as hourly, and by the peak discharge. Figure 3.18 is an example.

3.1.5 Special gaging methods

Variable backwater due to operation of a dam or other control on a stream may result, at times, in so little fall in the water surface through a reach that

POTOMAC RIVER BASIN 183

01600000 NORTH BRANCH POTOMAC RIVER AT PINTO, MD

LOCATION.--Lat 39°33'59", long 78°50'25", Mineral County, W. Va., Hydrologic Unit 02070002, on right bank at downstream side of Western Maryland Railway bridge at Pinto, 2.8 mi (4.5 km) downstream from Mill Run, and at mile 32.6 (52.5 km).

DRAINAGE AREA.--596 mi^2 (1,544 km^2).

WATER-DISCHARGE RECORDS

PERIOD OF RECORD.--October 1938 to current year.

REVISED RECORDS.--WSP 1332: 1943.

GAGE.--Water-stage recorder. Datum of gage is 648.23 ft (197.581 m) National Geodetic Vertical Datum of 1929. Prior to Dec. 10, 1938, nonrecording gage at highway bridge 250 ft (76 m) downstream at same datum.

REMARKS.--Water-discharge records good except those for winter periods, which are fair. Some regulation at low flow by Stony River Reservoir, 66 mi (106 km) above station (see station 01595200), and since December 1950, by Savage River Reservoir, 25 mi (40 km) above station (see station 01597500).

AVERAGE DISCHARGE.--41 years, 886 ft^3/s (25.09 m^3/s), 20.19 in/yr (513 mm/yr), unadjusted.

EXTREMES FOR PERIOD OF RECORD.--Maximum discharge, 37,000 ft^3/s (1,050 m^3/s) Oct. 16, 1954, gage height, 23.23 ft (7.081 m); minimum, 31 ft^3/s (0.88 m^3/s) Dec. 18, 19, 1943, gage height, 1.37 ft (0.418 m), result of freezeup.

EXTREMES OUTSIDE PERIOD OF RECORD.--Flood of Mar. 29, 1924, reached a stage of about 24 ft (7.3 m), discharge, about 55,000 ft^3/s (1,560 m^3/s). Flood of Mar. 17, 1936, reached a stage of about 23.5 ft (7.16 m), from floodmarks, discharge, about 50,000 ft^3/s (1,420 m^3/s).

EXTREMES FOR CURRENT YEAR.--Maximum discharge, 12,800 ft^3/s (362 m^3/s) Feb. 26, gage height, 13.28 ft (4.048 m); minimum, 126 ft^3/s (3.57 m^3/s) Nov. 15, gage height, 1.88 ft (0.573 m).

DISCHARGE, IN CUBIC FEET PER SECOND, WATER YEAR OCTOBER 1978 TO SEPTEMBER 1979
MEAN VALUES

DAY	OCT	NOV	DEC	JAN	FEB	MAR	APR	MAY	JUN	JUL	AUG	SEP
1	171	177	695	1750	603	4580	888	1250	1030	319	217	564
2	172	170	566	3990	483	4200	2160	1080	899	250	188	690
3	171	169	514	3870	482	4140	2070	964	770	241	172	394
4	165	169	2470	2700	509	6330	2220	1160	841	298	161	311
5	173	169	3050	1930	378	9570	2650	1540	769	1140	157	292
6	174	167	2240	1340	320	10400	2210	1470	586	631	152	3550
7	171	165	1500	1230	380	7890	1810	1290	498	376	147	2550
8	169	164	1540	2050	369	6350	1560	1130	502	286	151	1590
9	168	164	3720	1990	352	5480	1480	984	637	243	159	1180
10	172	163	3050	1610	343	4670	1500	893	1170	230	181	1010
11	171	163	1990	1190	340	3180	1310	1140	988	233	175	954
12	169	164	2090	908	350	2770	1180	1800	774	224	404	767
13	166	167	1740	835	370	2350	1090	2890	600	206	911	663
14	173	165	1260	790	390	2010	1060	2260	476	270	415	688
15	173	152	918	755	450	2150	1060	1720	403	260	266	735
16	179	250	735	739	1000	1750	1050	1430	357	471	232	724
17	184	618	711	897	1500	1580	1040	1140	327	448	205	604
18	183	554	698	1600	1100	1430	937	853	318	333	179	369
19	193	394	610	969	800	1230	845	758	320	280	231	338
20	177	299	581	871	750	1060	770	703	275	287	224	336
21	173	255	3330	2470	850	1010	717	631	236	300	217	368
22	169	231	3050	3340	1900	936	663	595	243	255	191	1470
23	167	224	2060	2310	2180	879	660	625	245	247	180	1250
24	161	226	1760	2100	5390	1160	895	1210	231	229	164	973
25	162	213	2680	2210	9280	2720	753	3080	215	228	157	839
26	165	209	2330	1530	11100	2200	746	3090	191	366	194	794
27	169	217	2200	1310	7550	2040	1530	2160	172	215	443	729
28	235	463	1670	1200	5220	1830	1880	1590	161	195	509	752
29	242	1090	959	1010	---	1530	1490	1570	180	210	572	800
30	201	678	1030	832	---	968	1400	1310	253	273	330	669
31	183	---	1030	969	---	877	---	1100	---	284	280	---
TOTAL	5501	8409	52777	51295	54739	99270	39624	43416	14667	9828	8164	26953
MEAN	177	280	1702	1655	1955	3202	1321	1401	489	317	263	898
MAX	242	1090	3720	3990	11100	10400	2650	3090	1170	1140	911	3550
MIN	161	152	514	739	320	877	660	595	161	195	147	292

CAL YR 1978 TOTAL 390628 MEAN 1070 MAX 10700 MIN 131 CFSM 1.80 IN 24.38
WTR YR 1979 TOTAL 414643 MEAN 1136 MAX 11100 MIN 147 CFSM 1.91 IN 25.88

Fig. 3.16. Published annual streamflow record.

the fall cannot be measured closely enough (or is affected by wind) for use in a stage-fall-discharge relation. Several alternatives are available.

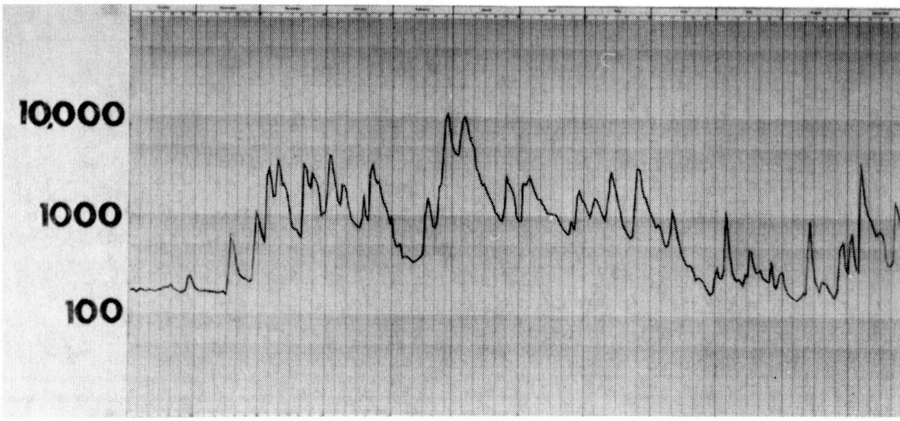

Fig. 3.17. Hydrograph of daily discharges of Figure 3.16.

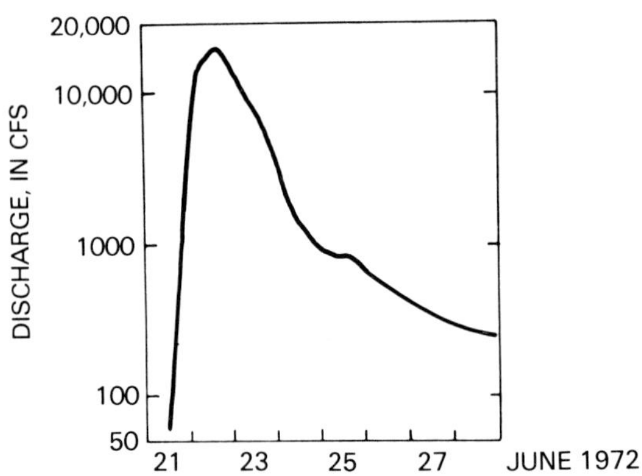

Fig. 3.18. Flood hydrograph.

(i) Deflection meter. The simplest alternative is use of a deflection vane set at a fixed position below the minimum stage and at a right angle to the current (Fig. 3.19). The angular deflection of the vane is a function of the velocity in the cross section, and the stream stage is related to the cross sectional area. Deflection and stage are measured at the gage site and the relation with discharge is based on current-meter measurements.

The deflection vane is used in small channels having little change in stage. Both downstream and upstream flow can be measured so it is suitable for tidal

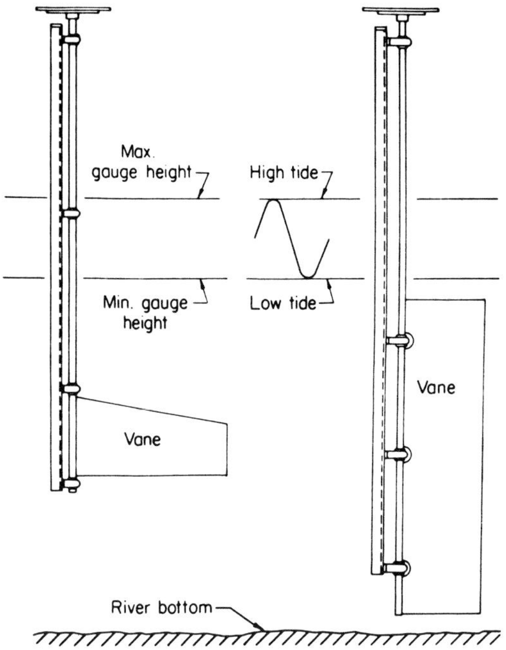

Fig. 3.19 Deflection-meter vanes.

channels. Discharges computed by this method are generally of low accuracy, especially those near zero.

(ii) Acoustic method. Velocity at some fixed level in the stream is obtained by determining the travel times of sound impulses moving in both directions along a diagonal path between transducers (sound generators or receivers) mounted near each bank. The layout of an operating system on Columbia River is shown in Figure 3.20. The installation and operation of that system is described by Smith and others (1971). Calibration is by current meter. Laenen and Smith (1982) have assembled published and unpublished information on the operation, applications, performance, and limitations of acoustic velocity measurement systems with specific applications to measurement of streamflow.

(iii) Electromagnetic method. Point velocity can be measured by an electromagnetic meter. Continuous records of velocity at one point in a cross section and of the stage can be transformed into a continuous discharge record after calibration with current-meter measurements. The reliability of the computed discharge depends on how closely the velocity at the one point represents mean velocity through the range of discharge experienced as well as on the reliability of the point velocities. Such an installation should be of relatively low cost but the electromagnetic meter is always subject to damage from

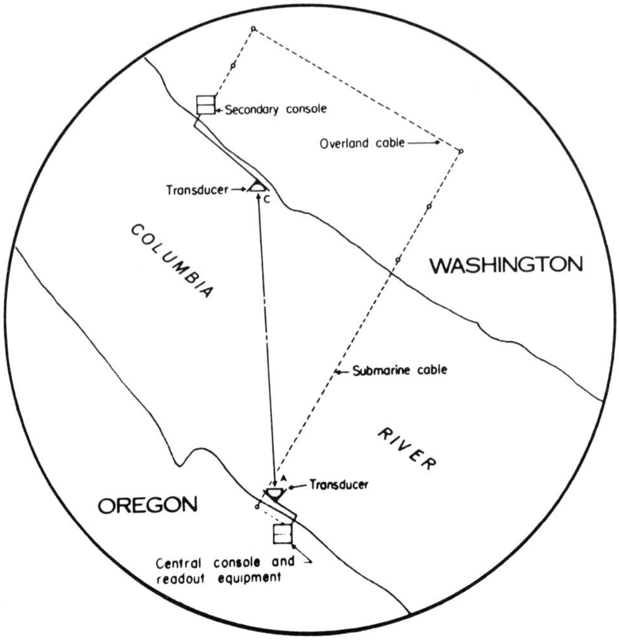

Fig. 3.20. Layout of an acoustic velocity measuring system (Smith and others, 1971).

debris if it is suitably located. Lack of stability of the meter calibration is also a problem.

Green and Herschy (1978) describe an electromagnetic gaging station based on the total-flow electromagnetic method. This application is still under development.

(iv) Dilution methods. Dilution methods of measuring stream discharge are useful under flow conditions that exist in shallow, extremely rough channels. Current-meter measurements are generally quicker and more reliable at all but extremely unfavorable sites.

The field procedure involves injecting a tracer of given concentration into a stream and sampling the concentration downstream where the tracer is completely mixed with the water. The tracer may be injected as a slug or at a constant rate. Useful tracers include salts, radioactive materials, and fluorescent dyes. The method requires skilled operators, considerable time, and a constant discharge.

Stream discharge for constant-rate injection is

$$Q = \frac{C_1 - C_2}{C_2 - C_b} q$$

where q is the tracer injection rate and C_b, C_1 and C_2 are concentrations of the stream at the injection point, of the injected tracer, and of the stream at the downstream cross section respectively. See White (1978) or Kilpatrick and Cobb (1984) for details.

3.1.6 Indirect measurements

It is often impossible to measure flood discharges at particular times because of lack of access to the site, shortage of manpower, or inadequate advance notice of the flood. Engineers have therefore devised methods of computing peak discharge after the passage of the flood; the common ones are slope-area, contracted-opening, flow-over-dam, and flow-through-culvert. These methods are based on hydraulic equations that relate the discharge to the water-surface profile, the geometry of the channel, and the channel roughness. These measurements may be expensive and they are less accurate than current-meter measurements.

The slope-area method is the most widely used. An ideal site is a reach of uniform channel on which the flood peak profile is defined on both banks by high-water marks. Surveys of these profiles and of channel cross sections, and estimates of the roughness coefficient in the Manning equation are required. Field experience in selecting roughness coefficients is desirable but guidance can be obtained from the photographs in the report by Barnes (1967).

Indirect methods are described by Barnes and Davidian (1978). Users guides to the various techniques are given by Dalrymple and Benson (1967), Bodhaine (1968), Matthai (1968), and Hulsing (1968).

Common to all these methods is the need to select the roughness coefficient subjectively. The roughness coefficient of a natural channel is a function of bed roughness, bank irregularity, effect of vegetation (if any), depth of water, channel slope, and other factors. No objective way of combining all these effects into one coefficient is available. Furthermore the verified value of a roughness coefficient for a particular channel reach and flood is affected by inaccuracies in measuring other variables in the Manning equation; this results in some apparent inconsistencies among verified roughness coefficients. To avoid this subjectivity, Riggs (1976) developed a simplified slope-area method in which discharge is related to cross-sectional area and to water-surface slope. A roughness coefficient is not used because, in natural channels, roughness and slope are related. The equation is

$$\log Q = 0.366 + 1.33 \log A + 0.05 \log S - 0.056 (\log S)^2$$

where Q is in cfs, A is average cross sectional area in square feet, and S is dimensionless slope of the water surface through the reach.

Mud flows or debris flows in small, steep-gradient channels leave physical evidence that may be misinterpreted as indicating the passage of a flood. A flood-peak discharge computed from such evidence might be extremely large, possibly greater than the potential for the basin or region. Costa and Jarrett (1981) describe how to make the proper interpretation of the evidence.

3.1.7 Crest-stage gaging stations

A crest-stage gaging station provides a record of peak stages and the corresponding discharges at a site. Its purpose is to provide data for defining the flood-peak frequency characteristics where the cost of collecting a continuous streamflow record cannot be justified.

A crest-stage gage usually consists of a 2-inch pipe mounted vertically on the stream bank. The pipe is capped at both ends and contains a wooden staff, and in the bottom cap, some ground cork. Intake holes in the bottom cap, and an air hole in the top cap, permit water to enter the pipe as the stream rises. The cork floats on the water and some of it sticks to the staff at the crest stage. The gage is inspected periodically to record the crest stage, to clean the staff, and to replenish the cork.

Crest-stage gages are usually located where a stage-discharge relation can be computed from channel characteristics, usually above a bridge or culvert, or on a reach suitable for step-backwater analysis. Figure 3.21 shows a crest-stage gage installation on an ephemeral stream in South Dakota.

Data produced by crest-stage gaging stations usually are limited to the maximum stage and discharge each year, although additional peaks may be recorded.

3.1.8 Time of travel

Time of travel is usually considered the mean travel time of water particles flowing from one cross section to another, at a given discharge. It is much longer than the time required for a flood wave to pass through the same reach.

Estimates of the rate of movement of waterborne particles in streams are needed for defining the waste-assimilative capacities of streams and to forecast the movement of a slug of contaminant such as might result from an accidental spill.

Various hydrologic tracers such as salt, radioisotopes, and fluorescent dyes have been used. The dye Rhodamine WT is a popular tracer. It is injected instantaneously at a stream cross section and the dye concentration is monitored as the dye cloud passes each of a series of cross sections. Dye concentration is measured by a fluorometer which is sensitive to concentrations as low as 0.05 parts per billion. Figure 3.22 shows the measured concentration at four points on the Mississippi River resulting from the injection of dye at Baton Rouge. The maximum concentration decreases and the longitudinal dispersion increases with distance downstream. Time of travel can be expressed as time for the leading edge, the peak concentration, or the last detectable dye to pass a given

Fig. 3.21. Crest-stage gage installation.

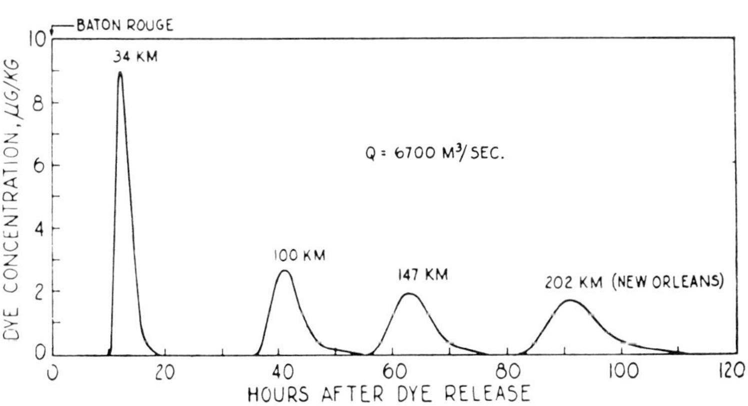

Fig. 3.22. Distribution of dye concentration with time at midstream sampling points, Mississippi River, Louisiana, September, 1965 (From Wilson, 1968).

point. All three times may be needed for some problems. Details of time-of-travel measurements are given by Hubbard and others (1982). See also White (1978) on dilution gauging.

Time of travel is greatly increased as stream discharge is decreased. Buchanan (1964) showed that tripling the discharge of Swatara Creek, Pa. reduced the travel time about half. Table 3.1 shows travel times through a 14.4 mi reach of St. Mary's River, Indiana for a range in discharge, as given by Eikenberry and Davis (1976).

TABLE 3.1

Travel time versus discharge

Discharge		Travel time (hours)	
cfs	% mean	leading edge	peak
22	4.6	105	130
120	25	33.0	40.0
620	130	13.8	16.8
810	170	12.0	14.8
1220	255	10.8	12.0

3.1.9 Sediment transport

Sediment-laden water is not only unsuitable for many uses without treatment, but it also deposits sediment in channels, canals, and reservoirs. Thus, designers and operators of water projects need information on the amounts and time distribution of sediment transported so as to minimize the detrimental effects.

Sediment is transported by a stream as suspended sediment, which is continually in suspension, and as bed load which moves by rolling, sliding, or bounding along the bottom. The amount of sediment being transported is highest during a period of flood runoff because of the erosion produced by the causative rainfall and because of the higher velocities and turbulence in the channels.

A sediment-discharge measurement, and a concurrent water-discharge measurement, produce data from which the following can be obtained: mean suspended sediment concentration, particle size distribution, specific gravity of the suspended sediment, temperature of the sediment-water mixture, water discharge, and the distribution of flow in the stream cross section.

In a sediment-discharge measurement water-sediment samples are collected at selected verticals in the cross section. The sample at each vertical is obtained as the sampler (Fig. 3.23) is lowered to the streambed and raised to the surface at a uniform rate or, alternatively, by collecting samples at selected points in the vertical. Laboratory analyses of the samples are used with the water-discharge measurement to produce the desired information. Details are given by Vanoni (1975, p. 317-349) and by Guy and Norman (1976).

On small, flashy streams nearly all the sediment is transported during the short periods of high discharge. Because these flood periods are rarely known

 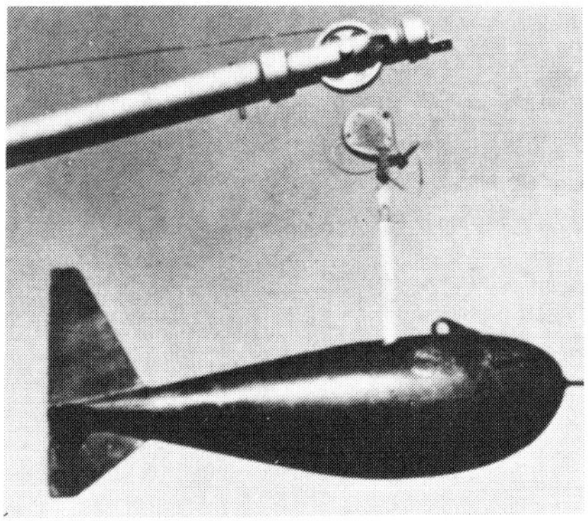

Fig. 3.23. Sediment sampler.

in advance, automatic sampling equipment has been developed. This may be a series of containers at different elevations for catching samples as the stream rises, or a pumping sampler programmed to take samples at selected intervals according to stream stage.

Records of sediment discharge over a considerable period of time are needed to define its variation with streamflow and with seasons, and to define the mean annual load of sediment transported.

If a daily record is desired, one or more depth-integrated samples are taken each day at one vertical; these are supplemented by periodic, more-detailed suspended-sediment measurements. The daily measured concentrations are plotted on a chart of gage height against time (usually the one from the analog recorder at the stream-gaging station) and the graph of sediment concentration is drawn between observed points using the stage graph as a guide. From the concentration graph and the streamflow record, the daily sediment load can be computed (Porterfield, 1972). Part of a published sediment record is shown in Figure 3.24.

The total suspended sediment load for a year can be approximated from occasional sediment measurements by use of a sediment-transport curve and a flow-duration curve. The former is a plot of sediment discharge against stream discharge (Fig. 3.25) and the latter shows the distribution of daily stream discharges during the year (Chapter 5). The duration curve is divided into intervals of time and the mean discharge for each interval is determined. This discharge is used to define the sediment transport for that interval of time. The annual sediment discharge is the sum for all of the intervals. See Colby (1963) and, for an example, Simmons (1976).

The measurements and the computed sediment loads described above are of suspended sediment. Total sediment load includes the bed load which is usually

11477000 EEL RIVER AT SCOTIA, CA--Continued

SUSPENDED-SEDIMENT DISCHARGE (TONS/DAY), WATER YEAR OCTOBER 1979 TO SEPTEMBER 1980

DAY	OCTOBER			NOVEMBER			DECEMBER		
	MEAN DISCHARGE (CFS)	MEAN CONCEN-TRATION (MG/L)	SEDIMENT DISCHARGE (TONS/DAY)	MEAN DISCHARGE (CFS)	MEAN CONCEN-TRATION (MG/L)	SEDIMENT DISCHARGE (TONS/DAY)	MEAN DISCHARGE (CFS)	MEAN CONCEN-TRATION (MG/L)	SEDIMENT DISCHARGE (TONS/DAY)
1	106	1	.29	1370	8	30	6180	52	868
2	100	1	.27	1340	9	33	5290	55	786
3	96	1	.26	5050	196	4510	4550	30	369
4	91	1	.25	11200	400	12100	3940	22	234
5	93	2	.50	9270	240	6010	3440	17	158
6	93	2	.50	14000	646	33500	3150	12	102
7	94	2	.51	29900	1260	104000	2900	10	78
8	94	3	.76	15700	340	14400	2680	14	101
9	96	3	.78	8960	140	3390	2500	8	54
10	96	3	.78	6130	70	1160	2310	10	62
11	96	3	.78	4630	45	563	2150	8	46
12	96	3	.78	3660	24	237	2020	6	33
13	99	3	.80	3040	16	131	1910	6	31
14	123	3	1.0	2610	11	78	1800	8	39
15	121	3	.98	2280	10	62	1730	5	23
16	133	1	.36	5220	271	6890	1650	4	18
17	156	2	.84	34700	1160	108000	1580	5	21
18	169	2	.91	20300	390	21400	1540	4	17
19	348	5	4.7	12100	160	5230	1920	10	52
20	984	20	53	8440	90	2050	2860	38	293
21	2400	85	624	6460	55	959	4540	127	1850
22	2540	68	466	7340	174	5370	8290	320	7160
23	2820	125	952	24800	970	67800	7980	220	4740
24	4720	234	3630	19500	502	28000	30300	976	84500
25	23000	1740	122000	25500	560	38600	37600	741	75800
26	20600	1150	64000	26800	643	48200	25700	355	24600
27	6580	220	3910	19000	295	15100	15900	215	9230
28	3700	70	699	12800	150	5180	11300	105	3200
29	2540	35	240	9510	94	2410	8680	60	1410
30	1930	16	83	7480	55	1110	9660	218	7070
31	1610	16	70	---	---	---	38300	1310	144000
TOTAL	75724	---	196743.1	359090	---	536503	254350	---	366945

Fig. 3.24. Published sediment record.

computed, or is estimated as a percentage of the suspended load. Measurements of bed load generally are not reliable. See Vanoni (1975).

If a stream flows through a reservoir most of the sediment will be trapped in the reservoir. The volume of material deposited and its volume-weight can be determined by surveys at intervals of several years. Of course the reservoir outflow will contain some suspended material; thus the total inflow load will be somewhat greater than that measured in the reservoir. Field measurement techniques to determine the volume occupied by sediments deposited in a reservoir are described in Vanoni (1975, p. 349-382). Runoff and sediment yield of ephemeral streams can be obtained from data collected at small reservoirs as described by Peterson (1962).

3.1.10 Chemical and biological quality

Whether water is considered of good or poor quality depends on the use to be made of it. Drinking water should not contain bacteria, suspended materials, certain minerals, or dissolved gases. Water for irrigation should contain only

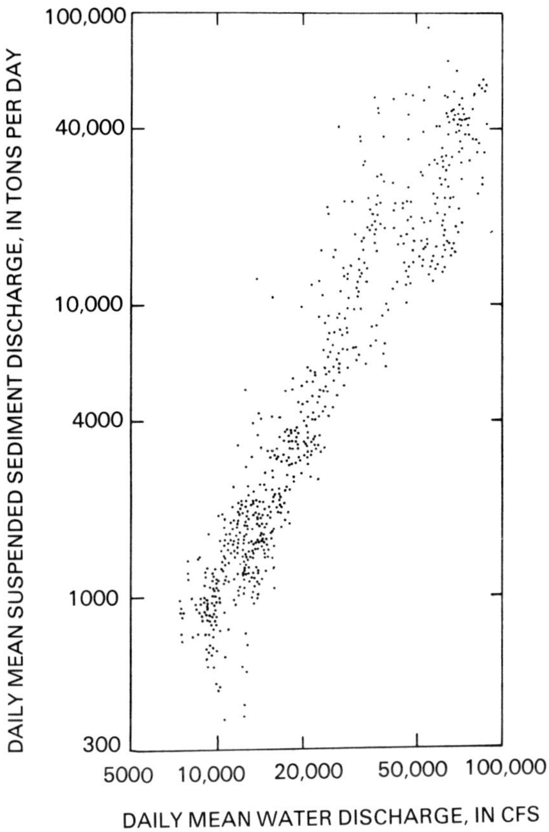

Fig. 3.25. Sediment-transport curve for Sacramento River at Sacramento, California (From Porterfield, 1980).

a limited amount of sodium and of some other elements; and suspended sediment is undesirable because it clogs the pipes and ditches. Some industrial processes require waters of very specific chemical content. Esthetically, water is considered good if it is clear (has little or no suspended or floating material), has no color or odor, and supports fish and other biota.

Water quality can be described in two ways, by identifying and quantifying the inorganic and the organic materials in the water, or by some measures of the effects of these materials. For example, the concentration of dissolved solids is related to the electrical conductance; the types of dissolved solids determine the pH, a measure of hydrogen-ion activity; and the concentration of dissolved oxygen is an indication of the biochemical condition of the water. These three indicators plus temperature can be measured in the field and are good general measures of water quality. But for certain uses one needs to know the kinds and concentrations of the various dissolved elements in the water, and whether dangerous bacterial or chemical pollutants are present. This information is obtained by laboratory analyses of samples of water from the stream.

Determination of water quality is a detailed and specialized operation as indicated by the wide range of physical, chemical, biological, and radiochemical information published in the annual water-data reports of the USGS for the National Stream-Quality Accounting (NASQUAN) stations. A purpose of these NASQUAN stations is to monitor changes in water quality, consequently specific conductance, pH, water temperature, and dissolved-oxygen concentration are recorded continuously. This detail is not necessary on natural (unpolluted) waters whose character does not change appreciably from year to year. Hem (1972, p. 40-50) describes how environmental influences affect natural water quality.

Methods for collecting and analyzing water-quality data are beyond the scope of this book. See Hem (1972, p. 60-68) for guidelines of sampling; Skougstad (1979) for determination of inorganic substances; Barnett and Mallory (1971) for determination of minor elements; Goerlitz and Brown (1972) for methods for analysis of organic substances; Greeson and others (1977) for methods for collection and analysis of aquatic biological and microbiological samples; Thatcher, Janzer, and Edwards (1977) for methods for determination of radioactive substances in water; and Stevens, Ficke, and Smoot (1975) for measurement of water temperature.

3.2 WEATHER OBSERVATIONS

The principal weather observations of concern to hydrologists are precipitation, temperature, and evaporation from water surfaces.

3.2.1 Precipitation

At most meteorological stations, the precipitation is caught in a can and the catch measured daily. The National Weather Service 8-inch nonrecording gage is shown in Figure 3.26. Details of the gage and instructions for making observations are given by the National Weather Service (NWS, 1972). When the precipitation is snow, the gage catch may not be representative. Then a sample of snow on the ground is obtained by inverting the can and cutting a vertical sample which is melted to determine the water content.

Recording precipitation gages commonly weigh the catch and record the cumulative catch on an analog chart or as punches at selected intervals on a paper tape. Data from these gages are commonly tabulated at hourly intervals; catches at shorter intervals can be obtained most readily from the digital tape, limited of course by the punch interval.

The tipping-bucket rain gage, commonly used in hydrologic studies, is actuated by small increments of rain (usually 0.01 inches in U.S.). The increments are recorded on an analog chart.

Fig. 3.26. National Weather Service nonrecording rain gage.

In remote areas, precipitation is caught in storage gages, 8-inch cans of considerable depth. The cans are charged with calcium chloride to melt snow and to prevent severe freezing of the catch. Oil is sometimes used to reduce evaporation between observations. The can is usually elevated on a tower and equipped with a shield to reduce wind velocity (Fig. 3.27). Storage gages are serviced at irregular intervals, sometimes only 3 or 4 times a year. Generally only seasonal or annual precipitation is obtained.

The catch of a precipitation gage depends on its location with respect to trees, buildings, and other obstructions. A location is considered satisfactory if there are no obstructions within an inverted 45-degree cone above the gage. However, a gage location may become unsuitable because of tree growth or building construction. Gages on windy, open areas tend to catch too little precipitation.

Although a precipitation gage may collect samples representative of the immediate locality, the record at the site may not describe the precipitation pattern some distance away, especially in mountainous country.

Natural precipitation may be affected by man's activities. Smoke and other airborne effluents from industrial areas tend to increase precipitation downwind, and large urban areas become hotter than undeveloped areas and induce more thunderstorms. In addition to inadvertent modifications, clouds are seeded to enhance precipitation in some regions. See Chapter 10.

3.2.2 Evaporation from water surfaces

Evaporation is commonly measured in an open pan. The widely-used Weather Bureau Class A pan is 4 ft in diameter and 10 inches deep (Fig. 3.28). It is

Fig. 3.27. Storage rain gage with shield.

Fig. 3.28. Evaporation pan.

filled to a depth of 8 inches and the depth is measured daily with a hook gage. Evaporation is the difference between readings, adjusted for any precipitation during the interval. When the water level has receded an inch, the pan is refilled. An evaporation station includes rain and temperature gages, and sometimes an anemometer for measuring wind.

The rate of evaporation from a pan is greater than that from a lake or reservoir because of the heat transferred through the pan walls. Although the "pan coefficient" to adjust annual pan evaporation to annual lake evaporation is considered to be about 0.7, for shorter periods it varies considerably and cannot be defined reliably.

Other ways of measuring water-surface evaporation include water budget, energy budget, and mass transfer. These methods require considerable data collected on a reservoir for a year or more.

Water budget is the simplest. Inflow, outflow, rainfall, and change in storage are measured. Water-surface evaporation can be computed because it is the only unknown in the water-budget equation. Reliability of the result depends on the magnitude of the evaporation relative to the magnitudes of the other elements; a small difference between two large numbers, each subject to some error, tends to be unreliable.

The energy-budget equation includes the energy inputs and outputs, all of which, except evaporation, can be measured. Data collection and analysis are described in a comprehensive interagency project report (U.S. Geol. Survey, 1954).

The mass-transfer equation is

$$E = N\mu\ (e_o - e_a)$$

where E is evaporation, N is the mass transfer coefficient, μ is wind speed, and e_o and e_a are saturation vapor pressure and actual vapor pressure respectively. Measurements of wind speed, and water and air temperature are required. The coefficient, N, must be derived from evaporation measured by another method (See Turner, 1966).

3.2.3 Temperature

Daily maximum and minimum air temperatures are obtained at many locations by the National Weather Service. Details of collection are given in their Observing Handbook No. 2 (NWS, 1972).

3.2.4 Snow accumulation

Snow fall is reported at weather stations but the accumulation of snow on the ground ordinarily is not. Snow on the ground is potential runoff which can be forecast if the amount and character of the snowpack are known. Measurement of the snowpack, called snow surveying, consists of measuring the depth and water content at snow courses (Fig. 3.29). A tube is used to extract cores of snow from the pack at designated distances along the line marking the snow course. Depth is recorded at each sampling point and the core is weighed to determine the water content (Fig. 3.30). Depths and water contents are averaged over all

Fig. 3.29. Snow surveyors at a marker designating one end of a snow course (U.S. Soil Conservation Service).

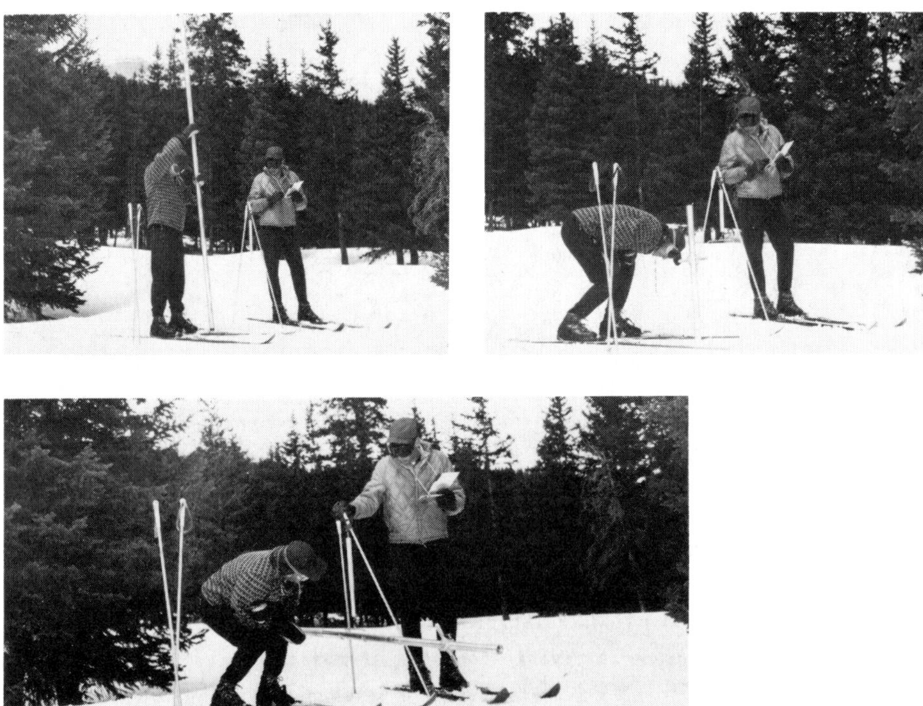

Fig. 3.30. Snow surveying: Inserting the tube, reading the depth, and weighing the tube with the snow core in it (U.S. Soil Conservation Service).

the sampling points to give the result. Snow surveys are usually made near the first of each of the spring months.

In mountainous regions travel to a snow course by skis or snowshoes is time consuming; if by helicopter the travel is expensive. The number of visits can be reduced by using a snow pressure pillow which is a flat flexible container filled with anti-freeze. The pillow indicates the snow-water equivalent by the pressure of the snow pack on the pillow. Pressure readings are transmitted by radio or satellite and transformed to water content by a previous calibration of the pillow (Rallison, 1981). Soil Conservation Service (1972) describes snow-surveying procedures in detail. Use of snow survey results is described in Chapter 11.

3.3 BASIN CHARACTERISTICS

Knowledge of drainage basin characteristics is useful in understanding a streamflow record and is a requirement in some methods of estimating flow characteristics at ungaged sites.

Size of drainage area is the most common basin characteristic. The drainage basin is delineated on topographic maps and its area measured. In arid or semiarid regions a major drainage basin may encompass an area which has no surface drainage; the contributing drainage area excludes that interior area.

Drainage area may not be a good indicator of streamflow if the topographic and the ground-water divides are not coincident. This disparity cannot be easily quantified but its recognition will help to understand the measured flows or to estimate flows at ungaged sites.

Basin topography influences runoff in several ways - steep slopes concentrate the rainfall quickly and result in high flood discharges; flat slopes result in slow runoff, increased ground-water recharge, increased evapotranspiration, and consequently in decreased total runoff. Basin topography is commonly quantified by some approximation of the slope of the main channel. An additional index is the percentage of lakes and swamps in the basin.

Vegetative cover can sometimes be quantified as the percentage of the area forested, or under cultivation. In a natural basin, the vegetative cover depends on climate and on soil characteristics.

Soil characteristics determine the rate of infiltration of rainfall to the soil and thus affect the rate and amount of runoff. The Soil Conservation Service (1971) classifies each soil into one of four ranges according to infiltration rate.

The geology of a basin affects the rate of runoff, the losses or gains along the channels, and especially the low-flow characteristics. Knowledge of the geology as it affects the water resource may be very useful even though it is only qualitative.

Precipitation usually is considered a basin characteristic. It can be expressed as mean annual precipitation, some measure of storm intensity such as the 12-hour storm precipitation at 50-year recurrence interval, or in other ways appropriate to a particular problem. Likewise, monthly mean temperatures are basin characteristics in that they are indicators of potential evaporation and transpiration rates and of whether precipitation will be snow and whether ice will form in streams.

Stream channel geometry, esthetic character, stability, and suitability for fish and wildlife habitat are useful descriptors although quantification is somewhat subjective.

3.4 TRANSMISSION OF HYDROLOGIC DATA

Conventionally, the stage data observed or recorded at a gaging station is obtained when the hydrographer visits the station, at monthly or longer intervals. This frequency of collection is adequate if the data are to be used for water-resource studies or for project design. But for water management purposes the data may be needed immediately, or in so-called "real time."

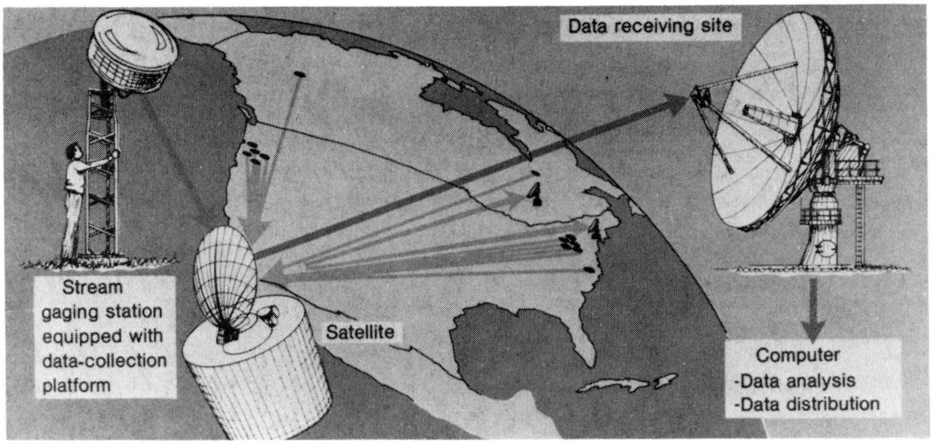

Fig. 3.31. Basic elements for a satellite data-collection system (From U.S. Geological Survey).

Various systems for transmitting data daily or more frequently utilize phone lines or radios. In the last few years automated satellite telemetry has become a practical means of providing water-data users with hydrologic information within the time frame needed. Satellite data-collection systems use earth-orbiting satellites to relay data from collection sites to receiving stations. The system consists of sensors, small radios called data-collection platforms, satellites, earth receiving sites, and a data processing and distribution system. See Shope and Paulson (1981) and Flanders (1981). The system is illustrated in Figure 3.31.

REFERENCES

Bailey, J.F. and Ray, H.A., 1966, Definition of stage-discharge relation in natural channels by step-backwater analysis: U.S. Geol. Survey Water-Supply Paper 1869-A, 24 p.

Barnes, H.H., Jr., 1967, Roughness characteristics of natural channels: U.S. Geol. Survey Water-Supply Paper 1849, 213 p.

Barnes, H.H., Jr. and Davidian, J., 1978, Indirect methods in Hydrometry, R.W. Herschy, ed., New York, John Wiley and Sons.

Barnett, P.R. and Mallory, E.C., Jr., 1971, Determination of minor elements in water by emission spectroscopy: U.S. Geol. Survey Techniques of Water-Resources Investigations, Book 5, Chapter A2, 31 p.

Bodhaine, G.L., 1968, Measurement of peak discharge at culverts by indirect methods: U.S. Geol. Survey Techniques of Water-Resources Investigations, Book 3, Chapter A3, 60 p.

Buchanan, T.J., 1964, Time of travel of soluble contaminants in streams: Jour. of Sanitary Engineering Division, ASCE, Vol. 90, No. SA3, 12 p.

Buchanan, T.J. and Somers, W.P., 1969, Discharge measurements at gaging stations: U.S. Geol. Survey Techniques of Water-Resources Investigations, Book 3, Chapter A8, 65 p.

Colby, B.R., 1963, Fluvial sediments - A summary of source, transportation, deposition, and measurement of sediment discharge: U.S. Geol. Survey Bulletin 1181-A, 47 p.

Costa, J.E. and Jarrett, R.D., 1981, Debris flows in small mountain stream channels of Colorado and their hydrologic implications: Bulletin of the Assoc. of Engineering Geologists, Vol. xviii, No. 3, August 1981.

Dalrymple, T. and Benson, M.A., 1967, Measurement of peak discharge by the slope-area method: U.S. Geol. Survey Techniques of Water-Resources Investigations, Book 3, Chapter A2, 12 p.

Davidian, J., 1984, Computation of water-surface profiles: U.S. Geol. Survey Techniques of Water-Resources Investigations, Book 3, Chapt. A15.

Eikenberry, S.E. and Davis, L.G., 1976, A technique for estimating time of travel of water in Indiana streams: U.S. Geol. Survey Water Resources Investigations 9-76, 39 p.

Flanders, A.F., 1981, Hydrological data transmission: World Meteorological Organization, Operational Hydrology Report No. 14, WMO-No. 559, 34 p., Geneva, Switzerland.

Goerlitz, D.F. and Brown, E., 1972, Methods for analysis of organic substances in water: U.S. Geol. Survey Techniques of Water-Resources Investigations, Book 5, Chapter A3, 40 p.

Green, M.J. and Herschy, R.W., 1978, New methods in Hydrometry, R.W. Herschy, ed., Chichester, UK, John Wiley and Sons.

Greeson, P.E., Elke, T.A., Irwin, G.A., Lium, B.W., and Slack, K.V., 1977, Methods for collection and analysis of aquatic biological and microbiological samples: U.S. Geol. Survey Techniques of Water Resources Investigations, Book 5, Chapter A4, 332 p.

Guy, H.P. and Norman, V.W., 1976, Field methods for measurement of fluvial sediment: U.S. Geol. Survey Techniques of Water-Resources Investigations, Book 3, Chapter C2, 59 p.

Hem, J.D., 1972, Study and interpretation of the chemical characteristics of natural water: U.S. Geol. Survey Water-Supply Paper 1473, Second Edition, 363 p.

Hubbard, E.F., Kilpatrick, F.A., Martens, L.A., and Wilson, J.F., Jr., 1982, Measurement of time of travel and dispersion in streams by dye tracing: U.S. Geol. Survey Techniques of Water-Resources Investigations, Book 3, Chapt. A9, 44 p.

Hulsing, H., 1968, Measurement of peak discharge at dams by indirect methods: U.S. Geol. Survey Techniques of Water-Resources Investigations, Book 3, Chapter A5, 29 p.

Kennedy, E.J., 1983, Discharge ratings at gaging stations: U.S. Geol. Survey Techniques of Water-Resources Investigations, Book 3, Chapt. A10.

Kilpatrick, F.A. and Cobb, E.D., 1984, Measurement of discharge using tracers: U.S. Geol. Survey Open-File Rept. 84-136, 73 p.

Laenen, A. and Smith, W., 1982, Acoustic systems for the measurement of streamflow: U.S. Geol. Survey Open-File Rept. 82-329, 45 p.

Matthai, H.F., 1968, Measurement of peak discharge at width contractions by indirect methods: U.S. Geol. Survey Techniques of Water-Resources Investigations, Book 3, Chapter A4, 44 p.

NWS, 1972, Substation observations, National Weather Service Observing Handbook No. 2: National Weather Service, Data Aquisition Division, Office of Meteorological Operations, Silver Spring, Md.

Peterson, H.V., 1962, Hydrology of small watersheds in western States: U.S. Geol. Survey Water Supply Paper 1475-I, 137 p.

Porterfield, G., 1972, Computation of fluvial-sediment discharge: U.S. Geol. Survey Techniques of Water-Resources Investigations, Book 3, Chapter C3, 66 p.

Rallison, R.E., 1981, Automated system for collecting snow and related hydrological data in mountains of western United States: Hydrological Sciences Bulletin, Vol. 26, No. 1, March 1981, p 83-89.

Rantz, S.E., 1982, Measurement and computation of streamflow: U.S. Geol. Survey Water-Supply Paper 2175, 631 p.

Riggs, H.C., 1976, A simplified slope-area method for estimating flood discharges in natural channels: U.S. Geol. Survey Jour. of Research, 4 (3), p 285-291.

Shope, W.G. and Paulson, R.W., 1981, Data collection via satellite for water management: Transportation Engineering Journal, ASCE, Vol. 107, No. TE4, July 1981, p 445-455.

Simmons, C.E., 1976, Sediment characteristics of streams in the eastern Piedmont and western Coastal Plain regions of North Carolina: U.S. Geol. Survey Water-Supply Paper 1798-O, p 10-14.

Skougstad, M.W., Ed., 1979, Methods for determination of inorganic substances in water and fluvial sediments: U.S. Geol. Survey Techniques of Water-Resources Investigations, Book 5, Chapter A1, 626 p.

Smith, W., Hubbard, L.L., and Laenen, A., 1971, The acoustic streamflow-measuring system on Columbia River at The Dalles, Oregon: U.S. Geol. Survey Open-File Report, Portland, Oregon.

Smoot, G.F. and Novak, C.E., 1969, Measurement of discharge by the moving-boat method: U.S. Geol. Survey Techniques of Water-Resources Investigations, Book 3, Chapter A11, 22 p.

Soil Conservation Service, 1971, Hydrology, SCS National Engineering Handbook, Section 4: Soil Conservation Service, U.S. Dept. of Agriculture, Washington, D.C.

Soil Conservation Service, 1972, Snow survey and water-supply forecasting: SCS National Engineering Handbook, Section 22, Soil Conservation Service, U.S. Dept. of Agriculture, Washington, D.C.

Stevens, H.H., Jr., Ficke, J.F., and Smoot, G.F., 1975, Water temperature - influential factors, field measurement, and data presentation: U.S. Geol. Survey Techniques of Water-Resources Investigations, Book 1, Chapter D1, 65 p.

Thatcher, L.L., Janzer, V.J., and Edwards, K.W., 1977, Methods for determination of radioactive substances in water and fluvial sediments: U.S. Geol. Survey Techniques of Water-Resources Investigations, Book 5, Chapter A5, 95 p.

Turner, J.F., Jr., 1966, Evaporation study in a humid region, Lake Michie, North Carolina: U.S. Geol. Survey Prof. Paper 272-G.

U.S. Geological Survey, 1954, Water-loss investigations Volume 1 - Lake Hefner studies: U.S. Geol. Survey Prof. Paper 269.

Vanoni, V.A., Ed., 1975, Sedimentation engineering: ASCE Manual of Engineering Practice, 745 p.

White, K.E., 1978, Dilution methods in Hydrometry by R.W. Herschy, Ed.: New York, John Wiley and Sons.

Wilson, J.F., Jr., 1968, Time of travel measurements and other applications of dye tracing: International Assoc. of Scientific Hydrology Publ. No. 76, Bern, p 252-262.

WMO, 1980, Manual on stream gauging: World Meteorological Organization Operational Hydrology Report No. 13, WMO - No. 519, Vol. 2, Geneva, Switzerland.

Chapter 4

STATISTICS

4.1 INTRODUCTION

Some understanding of the principles of statistics is needed to analyze and interpret hydrologic data properly. In this chapter principles are described without use of higher mathematics but with the assumption that the reader has some familiarity with statistics terminology, computation procedures, and elementary probability. Coverage is limited to frequency curves, statistical inference, and regression and correlation.

Statistical interpretations are valid only to the extent that the data and the physical conditions conform to the assumptions required by the statistical method used. Many statistical procedures require the assumption that the data are normally distributed, but the types of data used in hydrology commonly are not normally distributed, and some have no probability distribution at all. The hydrologist must select procedures most nearly suitable to the character of his data and must interpret the results accordingly.

This chapter provides the basic statistical tools, both analytical and graphical, and concludes with a discussion of the characteristics of hydrologic data. Applications to specific problems are given in later chapters. Most of the material in this chapter is taken from USGS publications authored by Riggs.

4.2 FREQUENCY CURVES

4.2.1 Distributions

The concept of a population of objects having a distribution of sizes (or of some other characteristic) is basic to the statistical method. It is not possible to collect enough data to define a frequency distribution exactly, but the existence of a particular one can be proven to the desired degree of confidence by repeating an experiment many times.

Kendall (1952, p. 23) reported the results of a dice-tossing experiment in which 12 dice were tossed simultaneously and the number of sixes was recorded for each toss. The dice were tossed 4,096 times with the results shown in Table 4.1. Also shown are the relative frequencies computed from the experimental results and the theoretical relative frequencies computed from the binomial distribution. The close agreement between the theoretical and experimental frequencies indicates that the binomial distribution is applicable to this problem.

The binomial distribution is a discrete distribution, that is, it can take values only at specific points along a scale. In the dice-tossing experiment,

TABLE 4.1

Results of dice-tossing experiment (After Kendall, 1952)

No. of sixes	Frequency	Relative frequency	Theoretical relative frequency
0	447	0.109	0.112
1	1,145	.280	.269
2	1,181	.288	.296
3	796	.194	.197
4	380	.093	.089
5	115	.028	.029
6	24	.006	.007
7 and over	8	.002	.001
Total	4,096	1.00	1.00

it is possible to obtain an integer number of sixes only; there is no such thing as 5.5 or 3.2 sixes.

More commonly, a variable may take any value along a scale. Such a variable and its distribution are known as continuous. A variable may be classified as continuous if it can take any value along a scale even though the limitations of measurement restrict the observations to discrete values. This condition exists with most natural phenomena.

To aid in understanding a distribution, consider 1,000 tree-ring indices ranging in size from 2 to 240. If these are grouped by six-unit increments of size, a histogram, or frequency distribution, is obtained (Fig. 4.1, left). The

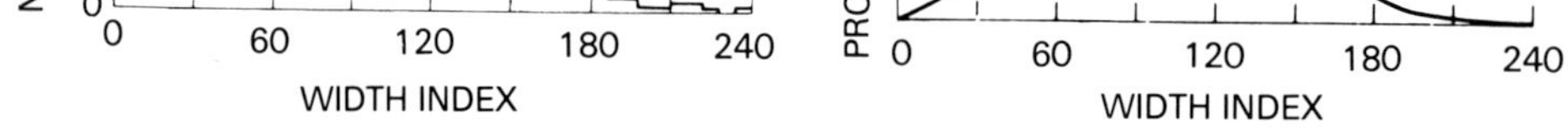

Fig. 4.1. Histogram and probability-density curve of 1000 tree-ring indices.

irregularity of the profile of this distribution is due to the small (in a statistical sense) number of indices used in its preparation. The greater the number used, the smoother would be the profile of the frequency distribution. If the number of observations approaches infinity and the size increment approaches zero, the enveloping line of the frequency distribution will approach a smooth curve. Then if the ordinate values are divided by a number such that the area under the curve becomes one, the resulting curve is a probability density curve, or probability distribution, as shown on the right of Figure 4.1. The process just described requires the additional assumption that the variable can take any value within the range, that the variable is continuous, not discrete.

A theoretical probability distribution describes the relation between size (or some other characteristic) and probability. For this relation to be valid, the individuals must occur randomly or be drawn randomly. The size of any individual drawn should not depend on the size of any one previously drawn. Probability, in the concept of frequency distributions, is defined as relative frequency. The distribution of the number of sixes obtained from repeated tosses of 12 dice could be illustrated by plotting the theoretical relative frequencies of Table 4.1. The relative frequencies of each of the 6-unit increments in Figure 4.1 could likewise be computed. In the first of these examples, a probability is associated with each possible outcome. In the second, a probability is associated with each increment of size; here the probability is of obtaining not a specific individual but any individual within the increment of size. This interpretation is required for continuous distributions because there is an infinite number of possible values and, thus, no probability of occurrence of a particular individual. Referring to the right curve of Figure 4.1, probability is related to the continuous distribution in the following way: The area under the curve represents the sum of all probabilities and therefore must equal one. Because every item was used in defining the distribution for which the total area is one, then the probability that any item will fall in the distribution is one and the probability that an item will fall in any segment of the distribution is the ratio of the area of that segment to the total area.

The distributions just described, both discrete and continuous, are called relative-frequency distributions, probability distributions, or just distributions. However, the probability interpretation is valid only if the data used are random. For example, the daily mean flow of a stream is closely related to the flows of previous days, so the distribution of daily means is not one to which the probability interpretation strictly applies. It is possible also to approximate a distribution which merely describes the sample. For instance, the distribution of grain sizes of a sample of a streambed is measured to characterize the material; there is no interest in the probability of obtaining a grain in a particular size range by additional sampling. Here the sample is not the individual grain but an aggregate of grains of various sizes.

Only a few standard theoretical distributions are widely used. Sampling theory and inference are based largely on the normal distribution with which the reader is assumed to be familiar. Other theoretical distributions will be introduced as appropriate.

4.2.2 Cumulative distributions

Suppose we know the probability density curve (probability distribution) for a variable and are interested in the probability of a random event being greater than some particular value E. This probability can be obtained by measuring or computing the proportion of the total area above the base value. For instance, the left curve of Figure 4.2 shows an area under the curve to the right of E of 0.1, that is, P = 0.1. Thus the probability of a random event exceeding E is 0.1.

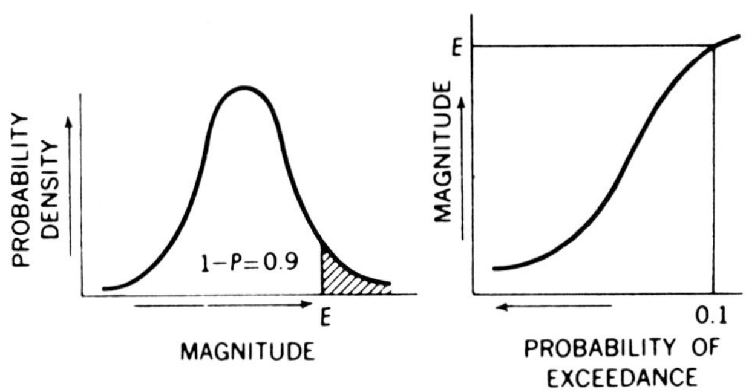

Fig. 4.2. Probability density curve (left) and its cumulative form (right).

Another form of the probability curve can be prepared by cumulating the probabilities from one end of the curve and plotting each of these cumulated probabilities against the magnitude of its appropriate event. The result is the right-hand curve of Figure 4.2. Cumulative distributions are commonly plotted to a probability scale such that the theoretical curve is a straight line. Such a scale can be devised for any two-parameter distribution. Normal probability plotting paper is widely known and used. Gumbel plotting paper is used in many hydrologic frequency analyses. (Although the Gumbel extreme-value distribution is a three-parameter distribution, one parameter, the skew, is constant for the form used and permits the construction of a scale which gives a straight-line plot). Both the normal and Gumbel probability plotting papers are available with either arithmetic or logarithmic ordinate scales. Thus plotting papers for four distributions - normal, log-normal, Gumbel, and log-Gumbel - are available.

When the probability density curve is cumulated from the right end, the probabilities of exceeding the various magnitudes are obtained. If cumulated from the left, probabilities of being less than those magnitudes are obtained. The appropriate cumulative curve, commonly called a frequency curve in hydrology, depends on the desired use.

A more detailed examination of the relation between a probability density curve and its cumulative form (the frequency curve) follows. We begin with the two normal distributions shown in Figure 4.3. Their cumulative forms can be

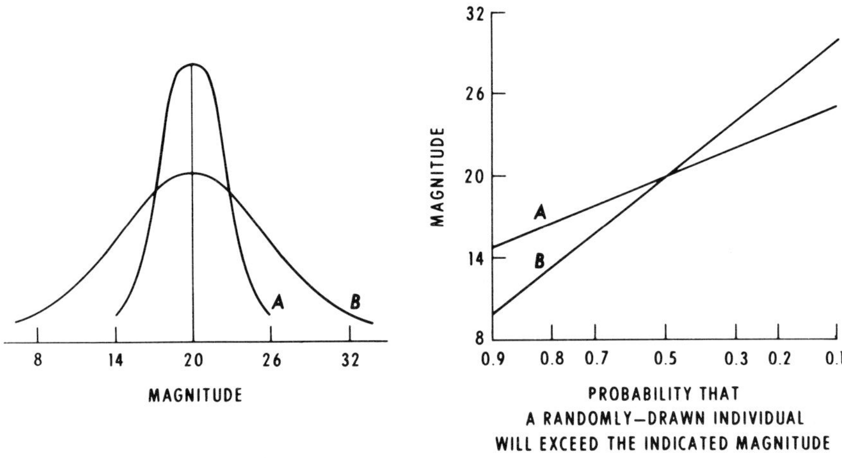

Fig. 4.3. Two normal distributions and their cumulative forms.

expressed as straight lines by use of the special abscissa scale which is derived from the characteristics of the normal distribution. Both distributions have the same median value, 20, and these medians plot at 0.5 probability on the cumulative graph. The variability of a distribution is indicated by the slope of the cumulative distribution; that is, the greater the variability, the greater the slope. The standard deviation is half the difference between magnitudes at probabilities of 0.16 and 0.84; (See a table of the cumulative normal distribution).

Many frequency distributions are nonsymmetrical. For such distributions, the mean, median, and mode have different values which consequently correspond to different probabilities on the cumulative graph. A nonsymmetrical distribution is classified as skewed. Skewness may be shown graphically as right or left; it may be described mathematically by a number, either positive or negative. Two skewed distributions and a symmetrical distribution are shown in Figure 4.4, which also shows the corresponding cumulative distributions (frequency curves).

For a normal, or any symmetrical, distribution the mean and median are the same value. Thus, the value corresponding to the probability of 0.5 on the cumulative frequency curve is the mean as well as the median for such distributions. The relative positions of the mean, median, and mode for skewed

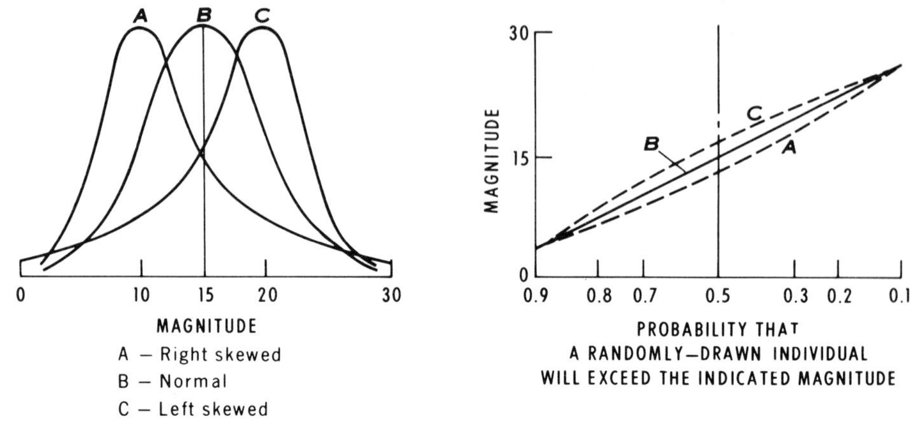

Fig. 4.4. Normal and skewed distributions and their cumulative forms on a normal-probability plot.

distributions are shown in Figure 4.5. Only the median value can be determined from the cumulative plot. The position of the mean with respect to the median

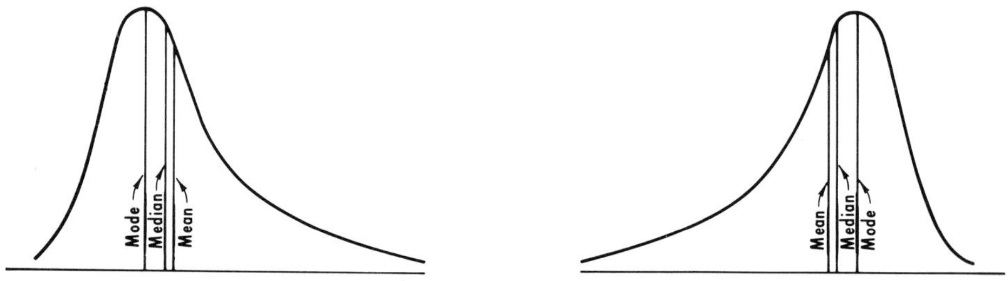

Fig. 4.5. Relative positions of the mean, median, and mode for positive or right-skewed (left graph) and for negative or left-skewed (right graph) distributions.

on the cumulative plot depends on the degree of skewness, the direction of skewness, and the direction in which the frequency distribution is cumulated. For example, 43 percent of the samples drawn from a particular right-skewed distribution will exceed the mean and 57 percent will be less than the mean. Thus, if the distribution is cumulated from the high end, the mean is to the right of the median; if cumulated from the low end, the mean is to the left of the median. These relations are reversed for left-skewed distributions. Figure 4.6 illustrates the relations. The probability scales of the two plots of

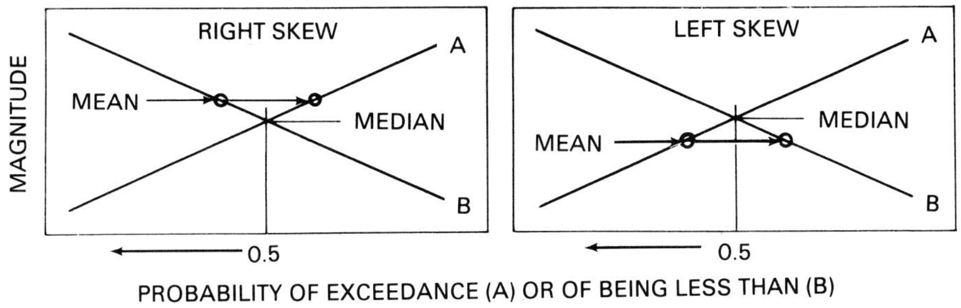

Fig. 4.6. Frequency curves showing effect of direction of skew and direction of cumulation on position of the mean with respect to the median.

Figure 4.6 are different; each is designed for the particular distribution plotted.

4.2.3 Recurrence interval

Frequency curves of hydrologic data commonly relate magnitude to recurrence interval or return period instead of to probability. Recurrence interval is the reciprocal of probability when the population consists of annual events. Thus the recurrence interval of a particular value is the average number of years between occurrences greater than (or less than) that value.

4.2.4 Graphical fitting

A frequency curve can be defined graphically from a list of items. In this method, each individual in the sample is assigned a probability or recurrence interval. Then magnitudes of the individuals are plotted against these probabilities or recurrence intervals, and a line is drawn to properly interpret the points.

Assignment of probabilities is by means of a plotting-position formula. Various formulas may be used, each based on a different assumption as to the characteristics of the sample. Langbein (1960) relates the better-known plotting-position formulas to their underlying assumptions. Benson (1962b) compares the results of using various plotting positions on the economics of engineering planning. A widely used formula is the Weibull

$$RI = 1/p = (n+1)/m$$

where RI is recurrence interval in years, p is probability of an exceedence in any one year, n is the number of items in the sample, and m is the order number of the individual in the sample array (Dalrymple, 1960). Although Beran (1981, p 26) justifies use of the Gringorten formula, the Weibull is used in this book because of its wide use in the United States. The sample data may be arrayed arranged in order of magnitude - beginning with the largest as No. 1, or with the smallest as No. 1, according to whether the frequency curve is to describe the probability of exceedence or of being less than. A distribution curve can be cumulated from either end, and in the graphical method this effect is accomplished by selecting the direction in which the data are arrayed.

The next step is plotting magnitude against recurrence interval (or probability) on a graph. If arithmetic coordinates are used, an S-curve usually results. It is difficult to define such a curve by the few observations; it is customary, therefore, to use a graph sheet having the abscissa graduated in such a way that a particular theoretical frequency curve will plot as a straight line. Such graph sheets are available for the normal, log-normal, and Gumbel Type I distributions. It is possible to prepare such a scale for any two-parameter distribution.

Although sets of data of the same type may not appear to lie on straight lines on a particular plotting paper, the lines of good fit usually are only slightly curved in one direction. An additional advantage of the probability graph appears when a straight line is a reasonable interpretation of the plotted points; then the straight line is a frequency curve of the theoretical type on which the plotting paper is based. It should be clearly understood that a frequency curve is not necessarily normal just because the points are plotted on normal-probability paper (or has a Gumbel distribution because the points are plotted on Gumbel probability paper); only when the frequency curve is a straight line is this true.

The mean of a normal distribution corresponds to the 0.5 probability or to the 2 year recurrence interval. But a curved line on normal probability paper represents a skewed distribution whose mean is not at 2-year recurrence interval. The effect of a skew on the relation of mean to recurrence interval is easily demonstrated by use of the Gumbel Type I distribution which has a fixed positive skew. As used for flood analyses, the mean occurs at 2.33-year recurrence interval. But if the same Gumbel distribution is used to represent the frequency of floods less than, the positions of the mean and median are reversed, and the mean plots at about 1.59 years. This effect is shown by Figures 4.5 and 4.6. The discharge corresponding to the 2.33-year recurrence interval as obtained from a curved line on Gumbel probability paper is not the mean. It can, however, be used as a characteristic discharge as could the 2-year value or any other near the central part of the distribution.

The annual discharges for the years 1915-50 inclusive in Table 4.2, column 2, can be used to define a frequency curve. The curve can be cumulated from the

TABLE 4.2

Computation of plotting positions

Water year	Discharge	Order number, m; highest as No. 1	Plotting position (n+1)/m	Order number, m; lowest as No. 1	Plotting position $(n+1)/m_1$
1915	264	34	1.09	3	11.2
16	374	11	3.37	26	1.43
17	332	19	1.95	18	2.06
18	346	16	2.31	21	1.76
19	359	13	2.85	24	1.54
1920	333	18	2.06	19	1.95
21	483	3	11.2	34	1.09
22	417	5	7.40	32	1.16
23	346	17	2.18	20	1.85
24	320	21	1.76	16	2.31
1925	271	31	1.19	6	6.16
26	214	36	1.03	1	37.0
27	530	2	18.5	35	1.06
28	304	25	1.48	12	3.09
29	271	32	1.16	5	7.40
1930	271	33	1.12	4	9.25
31	304	26	1.43	11	3.37
32	400	9	4.11	28	1.32
33	327	20	1.85	17	2.18
34	415	6	6.16	31	1.19
1935	402	8	4.62	29	1.28
36	362	12	3.09	25	1.48
37	320	22	1.68	15	2.47
38	272	30	1.23	7	5.30
39	244	35	1.06	2	18.5
1940	279	28	1.32	9	4.11
41	303	27	1.37	10	3.70
42	310	24	1.54	13	2.85
43	275	29	1.28	8	4.62
44	317	23	1.61	14	2.65
1945	350	15	2.47	22	1.68
46	387	10	3.70	27	1.37
47	359	14	2.65	23	1.61
48	449	4	9.25	33	1.12
49	406	7	5.30	30	1.23
1950	570	1	37.0	36	1.03

high end or from the low end, depending on whether the data are arrayed from the high end or from the low end. Both arrays are given in the table.

Arraying 20 or 25 items, that is, arranging them in order of magnitude, and assigning order numbers, can be done readily by observation. For a larger number of items, various schemes may be used.

The plotting positions given in Table 4.2 are preliminary recurrence intervals; probabilities would be their reciprocals; n is the number of items (1950 − 1915 + 1 = 36), and m is the order number.

Discharge is plotted against recurrence interval on a plotting paper having either an arithmetic or logarithmic ordinate scale and an abscissa scale based on the normal or some other 2-parameter distribution. Figures 4.7 and 4.8 are plotted from Table 4.2 data. Selection of the ordinate scale was arbitrary; both abscissa scales are based on the Gumbel Type I distribution. The lines are graphical interpretations. Neither of these curves represent Gumbel distributions because they are not straight lines.

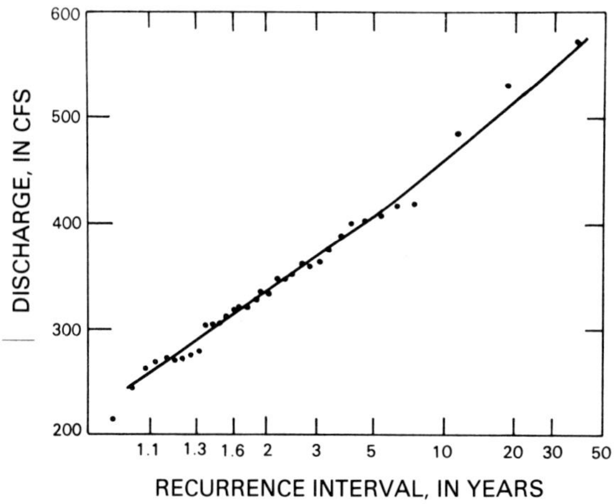

Fig. 4.7. Frequency curve based on data from Table 4.2 assuming that data are annual maximums.

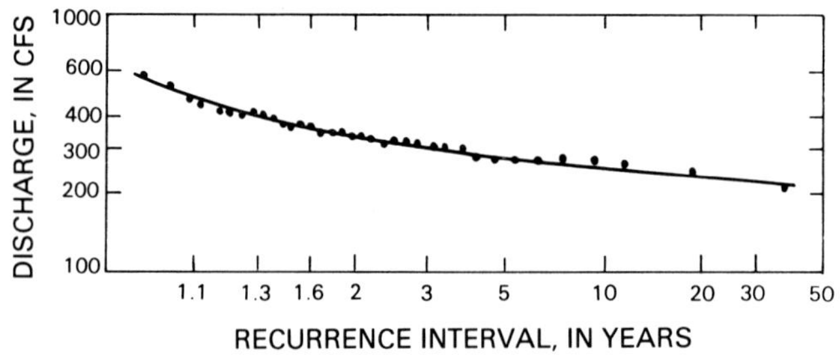

Fig. 4.8. Frequency curve based on data from Table 4.2 assuming that data are annual minimums.

4.2.5 Fitting theoretical distributions

The first step is to decide which distribution would be most appropriate. The ones most commonly used in hydrology are normal, lognormal, extreme value (Gumbel and Weibull), and log Pearson Type 3. Selection depends on the variable. Previous analyses may be used as a guide.

(i) Normal distribution. The normal distribution has only 2 parameters, mean and standard deviation. Estimates of these are computed from the discharges of Table 4.2 by

$$\text{Mean} = \bar{Q} = \Sigma Q/N = 12{,}486/36 = 347$$

$$\text{Variance} = S^2 = \frac{\Sigma Q^2 - (\Sigma Q)^2/N}{N-1} = \frac{4{,}542{,}500 - 155{,}900{,}196/36}{35} = 6055$$

Standard deviation = S = 77.8

Using these, the detailed characteristics of the distribution can be extracted from a table of the cumulative normal distribution, which is available in many statistics texts. In a table of the cumulative normal distribution, magnitude is expressed as the mean plus or minus some multiple of the standard deviation. We assume that $\bar{Q}$ and S are equal to their respective population parameters μ and σ. Then for selected values of magnitude in terms of μ and σ we determine the probabilities from the table at enough points to define the frequency curve. Results using a mean of 347 and a standard deviation of 77.8 are shown in Table 4.3.

TABLE 4.3

Frequency characteristics of data from Table 4.2 assuming a normal distribution

X_a		$P(X<X_a)$	$P(X>X_a)$	Recurrence Interval of Exceedance
$\mu - 2.0\sigma$	= 191	0.02	0.98	1.02
$\mu - 1.5\sigma$	= 230	.07	.93	1.08
$\mu - 1.0\sigma$	= 269	.16	.84	1.19
$\mu - .8\sigma$	= 285	.21	.79	1.27
$\mu - .5\sigma$	= 308	.31	.69	1.45
$\mu - .2\sigma$	= 331	.42	.58	1.72
	= 347	.50	.50	2.00
$\mu + .2\sigma$	= 363	.58	.42	2.38
$\mu + .5\sigma$	= 386	.69	.31	3.22
$\mu + .8\sigma$	= 409	.79	.21	4.76
$\mu + 1.0\sigma$	= 425	.84	.16	6.25
$\mu + 1.5\sigma$	= 464	.93	.07	14.4
$\mu + 2.0\sigma$	= 503	.98	.02	50.0

The same results could have been obtained from a plot on normal probability paper. The mean is plotted at 0.5 probability, the standard deviation is plotted plus and minus from the mean at probabilities of 0.16 and 0.84, respectively, a straight line is drawn through the plotted points, and probabilities at selected levels are read from the line.

The lognormal distribution is a normal distribution of the logarithms of the data. Either Naperian or common logarithms can be used but common logarithms are most widely used in hydrology. In terms of the untransformed data the lognormal distribution has a range from zero to plus infinity and thus is positively skewed. The lognormal distribution is usually treated simply as a normal distribution of logarithms and can be fitted to data as shown in the above example by substituting log Q for Q in Table 4.2.

(ii) Type I Extreme-value distribution (Gumbel). This is a 2-parameter distribution having a constant skew of 1.139. The parameters are

$$u = \bar{X} - \bar{y}_N/\alpha \qquad \text{and} \qquad 1/\alpha = S/\sigma_N$$

where u is the mode, $1/\alpha$ is a scale parameter, $\bar{X}$ and S are the sample mean and standard deviation respectively, and $\bar{y}_N$ and σ_N are functions of N, the number of items in the sample. Values of $\bar{y}_N$ and σ_N for N from 8 to 1,000 are tabulated by Gumbel (1958, p. 228). Part of Gumbel's table is given in Table 4.4.

TABLE 4.4

Means and standard deviations of reduced extremes (from Gumbel, 1958)

N	$\bar{y}_N$	σ_N
10	0.4952	0.9497
12	.5035	.9833
14	.5100	1.0095
16	.5157	1.0316
18	.5202	1.0493
20	.52355	1.06283
25	.53086	1.09145
30	.53622	1.11238
40	.54362	1.14132
50	.54854	1.16066
60	.55208	1.17467
80	.55688	1.19382
100	.56002	1.20649
200	.56715	1.23598
500	.57240	1.25880
1000	.57450	1.26851

The mean and standard deviation of the sample are computed, $\bar{y}_N$ and σ_N are read from the table, u and $1/\alpha$ are computed from the above formulas, and the straight line

$X = u + y/\alpha$

is determined. On Gumbel probability graph the mean is plotted at 2.33 years and the approximate relation

$y = \ln RI$

can be used to locate another point on the straight line.

Following is a sample computation for annual floods on Columbia River near The Dalles, Oregon for 1858-1946. Data are from USGS WSP 1080.
Mean flood is 606,200 cfs $= \bar{X}$

Standard deviation, $S = \sqrt{(\Sigma X^2 - N\bar{X}^2)/N - 1} = 175{,}200.$

From Table 4.4 for N = 89,

$\bar{y}_N = .558$ and $\sigma_N = 1.20$

then

$1/\alpha = S/\sigma_N = 175{,}200/1.20 = 146{,}000$

and

$u = \bar{X} - \bar{y}_N/\alpha = 606{,}200 - (.558)(146{,}000) = 524{,}700.$

The equation of the straight line is

$X = u + y/\alpha = 524{,}700 + 146{,}000y.$

The relation $y = \ln T$ may be used to define plotting points for large recurrence intervals.

$y = \ln T = 2.303 \log T.$

For T = 50 years, y = 3.91, and

$X = 524{,}700 + 146{,}000(3.91) = 1{,}096{,}000$ cfs.

The straight line is defined on the graph of Figure 4.9 by the points X = 606,200 cfs at 2.33 years and 1,096,000 cfs at 50 years. The plotted points for the period 1858-1948 are shown to indicate the fit of the computed line.

Several analysts have concluded that the above fitting procedure recommended by Gumbel (1958) may produce biased estimates. See Lettenmaier and Burges (1982) and references given in their paper.

(iii) Pearson Type 3 distribution. The Pearson Type 3 distribution is a flexible distribution in three parameters with a limited range to the left and unlimited range to the right. Plotting paper is not available for this distribution because skewness varies. This distribution is commonly fitted to the

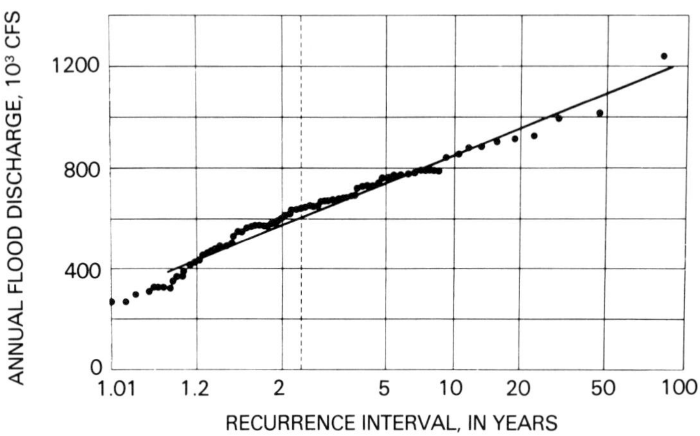

Fig. 4.9. Gumbel frequency curve.

common logarithms of flood magnitudes because this results in a smaller skew. The Pearson Type 3 distribution with zero skew is identical to the normal distribution.

Fitting a 3-parameter distribution starts with computation of the mean, standard deviation, and coefficient of skew. The mean and standard deviation (347 and 77.8) of the discharges of Table 4.2 were computed for fitting the normal distribution. The coefficient of skew is

$$C_s = \frac{N^2\Sigma Q^3 - 3N\ \Sigma Q\ \Sigma Q^2 + 2(\Sigma Q)^3}{N(N-1)(N-2)S^3}$$

$$C_s = \frac{(36)^2(1{,}738{,}665{,}756) - 3(36)(12{,}486)(4{,}542{,}500) + 2(12{,}486)^3}{(36)(35)(34)(6055)(77.8)}$$

C_s = 1.04.

These data parameters may be used to fit any 3-parameter distribution but the procedure is indirect because the probability density function cannot be integrated directly and thus there is no formula for the cumulative distribution. The relation between magnitude and probability of being larger than (or smaller than) is commonly determined from a table of frequency factors for the chosen distribution. Frequency factors for the Pearson Type 3 distribution, adapted from a more extensive table by Water Resources Council (1982), are given in Table 4.5. Specific points on the fitted distribution are computed by

$$X = \bar{X} + KS$$

where X is the variable at a selected recurrence interval (or probability), $\bar{X}$ is the mean, K is the frequency factor for the same recurrence interval, and S is the standard deviation.

Using the above value of mean, standard deviation, and skew, and Table 4.5, the discharge at 20-year recurrence interval would be

$$X = 347 + 1.88\ (77.8) = 493\ \text{cfs}$$

Additional points would be computed as needed to define the curve.

The log Pearson Type 3 distribution is fitted similarly except that the common logarithms of the discharges are used in computing the three parameters.

Frequency curve fitting is usually done by digital computer.

4.2.6 Evaluation of fitting methods

Fitting to theoretical distributions has the following advantages:

1. For the same theoretical distribution, every analyst using a given set of data would get the same answer.
2. Fitting can be done quickly by use of computer programs.
3. The curve can be completely described by 2 or 3 parameters and the name of the distribution.

and the following disadvantages:

1. Selection of the theoretical distribution is arbitrary.
2. No one theoretical distribution will adequately fit all hydrologic data of one type such as flood peaks.
3. Some data cannot be fitted by a distribution with only 3 parameters.
4. The highest and lowest data points may be given too much weight (the objective in hydrology is to best define one end of a frequency curve, not to define the best fit throughout the range).

Graphical fitting has the following advantages:

1. Procedure is simple and can be done quickly.
2. No assumption as to the type of distribution need be made.
3. Analyst can use other information in interpreting the plotted points.
4. Historical data may be readily incorporated.

and these disadvantages:

1. Even though the same plotting position formula is used, different analysts will draw somewhat different frequency curves.
2. A graphical curve cannot be described by 2 or 3 parameters.

TABLE 4.5

Frequency factors for Pearson Type 3 distributions

C_s	Recurrence Interval, in Years							
	100	20	10	3.33	2	1.43	1.11	1.05
2.0	3.61	2.00	1.30	.20	−.31	−.64	− .89	− .95
1.8	3.50	1.98	1.32	.24	−.28	−.64	− .94	−1.02
1.6	3.39	1.96	1.33	.28	−.25	−.64	− .99	−1.09
1.4	3.27	1.94	1.34	.31	−.23	−.64	−1.04	−1.17
1.2	3.15	1.91	1.34	.35	−.20	−.63	−1.09	−1.24
1.0	3.02	1.88	1.34	.38	−.16	−.62	−1.13	−1.32
.8	2.89	1.84	1.34	.41	−.13	−.60	−1.17	−1.39
.6	2.76	1.80	1.33	.44	−.10	−.59	−1.20	−1.47
.4	2.62	1.75	1.32	.47	−.07	−.57	−1.23	−1.52
.2	2.47	1.70	1.30	.50	−.03	−.55	−1.26	−1.59
0	2.33	1.64	1.28	.52	0	−.52	−1.28	−1.64
− .2	2.18	1.59	1.26	.55	.03	−.50	−1.30	−1.70
− .4	2.03	1.52	1.23	.57	.07	−.47	−1.32	−1.75
− .6	1.88	1.46	1.20	.59	.10	−.44	−1.33	−1.80
− .8	1.73	1.39	1.17	.60	.13	−.41	−1.34	−1.84
−1.0	1.59	1.32	1.13	.62	.16	−.38	−1.34	−1.88

4.2.7 Interpretation of frequency curves

A frequency curve based on random homogeneous data is an estimate of the cumulative probability distribution of the population from which the sample was drawn. The following interpretations of the frequency curve require the assumption that the curve is a good representation of the population distribution.

Referring to the graphical curve of Figure 4.7, the recurrence interval of 500 cfs is 16 years. This means that the annual maximum will exceed 500 cfs at intervals averaging 16 years in length, or that the probability of the annual maximum exceeding 500 cfs in any one year is 1/16.

From Figure 4.8 the recurrence interval of 250 cfs is 13 years. Thus, the annual minimum discharge will be less than 250 cfs at intervals averaging 13 years in length, and the probability that the minimum discharge in any one year will be less than 250 cfs is 1/13.

Many interpretations of frequency curves in hydrology have been stated in terms of the probability "of equaling or exceeding" a selected value. Most variables in hydrology, notably streamflow, are continuous - but reported as discrete - and the theoretical probability of occurrence of any particular value in a continuous distribution is zero. Therefore it seems desirable to delete "of equaling."

The interpretations of frequency curves given above will not answer questions such as the probability of an event of 10-year recurrence interval being exceeded in a 10-year period. Intuitively, one might expect that probability to be 0.5, but it is not. The correct probability can be computed as follows. Since the probability of exceeding the 10-year event in 1 year is 0.1, the probability of not exceeding it in 1 year is 0.9. Then, the probability of not exceeding it in 10 years is, by the multiplicative law of probabilities

$$(0.9)^{10} = 0.35$$

and the probability of one or more events exceeding the 10-year event in 10 years is 0.65. A more complete interpretation of a frequency curve is given by Riggs (1961).

Although frequency curves are used as though they were accurate representations of the population distribution, we know that they may not be. Benson (1960) sampled from a known distribution and showed a wide range in shape and position of frequency curves defined by different samples of the same size. Another way of assessing the reliability of a frequency curve is by computing the confidence limits. Chow (1964, p. 8-31) describes a method. These computations indicate that the frequency curve is most reliable in the vicinity of the mean.

4.2.8 Describing frequency characteristics

A mathematically-fitted frequency curve can be described by the name of the distribution and the mean, standard deviation, and skew. For use in hydrologic analyses, or for quantifying a regulatory flow, the discharge at some point on the frequency curve is used; for example the 100-year (0.01 probability) flood and the 10-year low flow are commonly used indices. The measure of central tendency commonly used is the median which is the discharge at 2-year recurrence interval or at 0.5 probability.

Variability of a distribution is described by its standard deviation, S. For comparison of variabilities among frequency curves the dimensionless ratio of standard deviation to the mean is used. This is called the coefficient of variation

$$C_v = S/\bar{X}$$

If logarithms of the data are used to define the frequency curve, the standard deviation of the logarithms is called the index of variation, I_v; unlike S, I_v is already dimensionless and need not (and should not) be divided by the mean.

The mean and standard deviation of graphically-fitted frequency curves are readily obtained from the graph if the frequency curve is a straight line on normal or lognormal probability paper. But the usual graphically-fitted frequency curve is not a straight line on the plotting paper used; consequently, the mean is not at a known probability or recurrence interval, the standard deviation cannot be accurately determined, and the curvature indicates the existence of skewness. For such curves it is customary to use the median flow as the characteristic of central tendency. Discharges at selected recurrence intervals also are readily obtained but an index of variability can only be approximated from the graphically-defined curve.

4.3 STATISTICAL INFERENCE

We have 54 years of record on the Rappahannock River of Virginia and might ask two questions about the mean flow. First, what is the mean flow for the period of record? This is a unique value which can easily be computed. The second question, what is the mean flow of the stream?, cannot be answered definitely. We can only assume that the mean of the 54-year sample is an estimate of the true (population) mean. In other words, we infer the population characteristics from those of a sample from that population.

Statistical inference is based on the theory of sampling. From a population of known characteristics many samples are drawn (either actually or conceptually), and the relation of the sample characteristics to the population characteristics is defined.

Sampling theory requires use of the concept of a probability distribution. Assume that the distribution of some random variable is normal with mean, μ, and standard deviation, σ, as shown in Figure 4.10. (The term "random", as used here, means that the probability of drawing any one item of the population is the same as for any other).

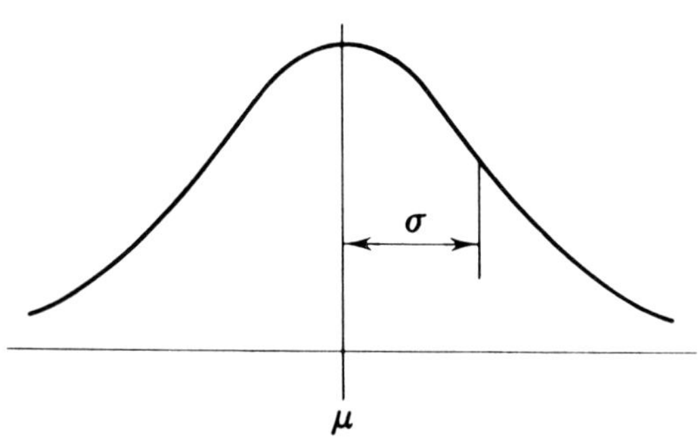

Fig. 4.10. Normal distribution.

Now suppose we take many samples of size N from this distribution, compute the mean of each of these samples, and compute the mean and variance of these sample means. The distribution of the means of samples of size N is superposed on the original distribution in Figure 4.11. It can be shown that the distribution of the means is centered at μ and that the standard deviation of the distribution of means is $\sigma/\sqrt{N}$. Therefore, the mean of the means of samples of

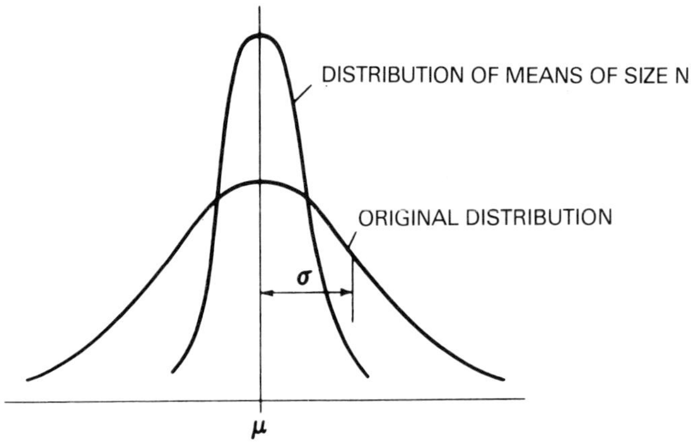

Fig. 4.11. Distribution of means of samples from a normal distribution.

size N is an unbiased estimate of μ. Furthermore, the mean of one sample is an unbiased estimate of μ. Consequently, we infer that the sample mean, $\bar{X}$, is an estimate of the population mean. Obviously, if we used other samples we would obtain different estimates of the population mean.

From a single sample, we can appraise the reliability of the estimate, $\bar{X}$, of the population mean. The distribution of means of values drawn from a normal distribution is normal. Consequently, two-thirds of the values should fall within one standard deviation ($\sigma/\sqrt{N}$) on each side of the mean. However, we do not know σ so we have to substitute S for it (where S is the standard deviation computed from the sample). The distribution of $\bar{X}$ having a standard deviation of $S/\sqrt{N}$ is known as the Student's t distribution, values of which are tabulated in statistics texts for various sizes of N.

Suppose now that we have K samples of size N and have defined K different sampling distributions of the mean of size N. For each sampling distribution, we can define a mean and a range of reliability, and we are interested in whether such a range includes the true mean μ. Considering the range as a random interval, we may state that the probability (P) that the random interval includes μ is 1−e, where e is the level of significance. Mathematically, for e = 0.32,

$$P[\bar{X} - \sigma/\sqrt{N}) < \mu < (\bar{X} + \sigma/\sqrt{N})] = 1-e = 0.68 .$$

The interval in brackets is called a confidence interval and the extremes are called confidence limits. Note that the above relation holds only if we use σ instead of S. If we use S, then 1−e is a function of the sample size, and the appropriate probability statement is

$$P\,[\bar{X} - tS/\sqrt{N}) < \mu < (\bar{X} + tS/\sqrt{N})] = 1-e = 0.68,$$

where Student's t is 1.09 for 10 degrees of freedom, for example. The width of the confidence interval increases as the level of significance decreases. For example, the 95 percent confidence limits (e = 0.05) are

$$(\bar{X} - 2.23S/\sqrt{N}) < \mu < (\bar{X} + 2.23S/\sqrt{N}),$$

where 2.23 is from the t table for 10 degrees of freedom.

The confidence interval described in the probability statement is a random interval, not a specific one. The probability statement (for e = 0.05) means that 95 percent of a large number of intervals similarly obtained would include the true mean. This probability statement cannot be extended to one specific interval because a specific interval either contains the true mean or it does not and the probability is either one or zero. The true mean is not a variable, it is unique.

But we are interested in making a probability statement about one specific interval. We may say that the probability of our obtaining a random interval which includes the true mean is 0.95, or that we have 95-percent confidence that the interval obtained includes the true mean.

Using the above theory, from a random sample we can compute an estimate of the population mean and a measure of its reliability. This is an example of statistical inference.

Returning to the sampling theory, consider the distribution of variances of samples of size N from a normal distribution. This distribution is not centered around σ^2 but is to the left of it, as shown by the left graph of Figure 4.12. Therefore S^2 is known as a biased estimator of σ^2. It can be made unbiased by multiplying it by N/(N−1) as shown in the right graph of Figure 4.12. The standard deviation of the sampling distribution of S^2 (N/N−1) can also be computed.

A further use of inference is in testing hypotheses. One example will be given. Suppose we set up the null hypothesis, H_0, that the mean of a population is zero; that is, we hypothesize that there is no difference statistically between the mean and zero. This null hypothesis is written

$$H_0{:}\mu = 0$$

We draw a sample from this population and compute the statistics of the sampling distribution of the mean. We need some estimate of the hypothetical sampling distribution of means, and so we define it as normal with mean zero, and standard deviation equal to $S/\sqrt{N}$ as computed from the sample (Fig. 4.13).

Now if $\bar{X}$ (as computed from the sample) lies within one standard deviation of zero, we would conclude that there is no basis for doubting the hypothesis.

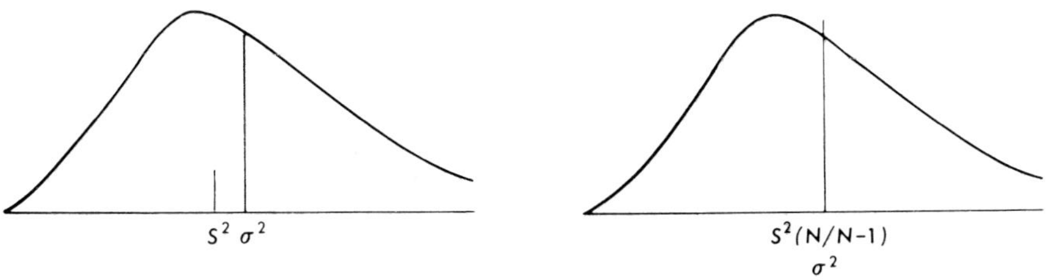

Fig. 4.12. Distribution of variances of samples.

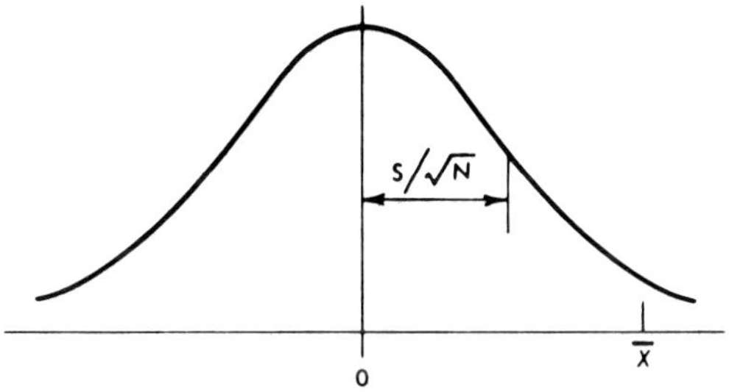

Fig. 4.13. Hypothetical sampling distribution of means.

If, on the other hand, $\bar{X}$ were two or three standard deviations away from zero, we would conclude that it is unlikely that the mean of the population is zero. For this latter condition the probability is small of obtaining an $\bar{X}$ of such size from a population having a mean of zero. Therefore, we would reject the hypothesis and would state that the result was significant at a certain probability level, meaning that the results obtained differ significantly from the hypothesis.

A common problem is the test of significance of a regression coefficient. The null hypothesis is again that the true value of the regression coefficient is zero, and the test may be made in the same way as before. See Test of Significance under Regression Methods.

4.4 CORRELATION AND REGRESSION

The distinctions between correlation and regression must be recognized in order to apply and interpret either of the methods. These distinctions are very marked although they may seem of little importance because of the similarity of the computation procedures. Dixon and Massey (1957, p. 189) made the following distinction between the two:

"A regression problem considers the frequency distribution of one variable when another is held fixed at each of several levels. A correlation problem considers the joint variation of two measurements, neither of which is restricted by the experiment."

Correlation is a process by which the degree of association between samples of two variables is defined. The correlation coefficient is a mathematical definition of that association. It is, of course, possible to compute a correlation coefficient from any two sets of data. The mathematical definition of association implies no cause-and-effect relation nor even that the relation between the two variables results from a common cause.

Correlation theory requires that the data be drawn randomly from a bivariate normal distribution. However, McDonald (1957) reported that experimental sampling studies show the nonnormality effects, usually regarded as disturbing by statisticians, to be of inconsequential magnitude geophysically. A further requirement of correlation is that both variables X and Y be without error due to measurement. Nothing can be measured without error, so the above requirement is one of degree. The question of the error allowable is subject to arbitrary decisions, particularly since the true error of the data is never known.

The end product of the process of correlation is the correlation coefficient; it is not an equation. The equations which describe Y as a function of X, and X as a function of Y, are regression equations, not correlation equations. Another way of stating the distinction between correlation and regression is that correlation measures the degree of association between two variables, whereas regression provides equations for estimating individual values of one variable from given values of the other.

Reliability of correlation results depends on the number of items used to compute the correlation coefficient and the magnitude of the computed correlation coefficient. Confidence limits are quite wide for samples of 30 items or less, unless the correlation coefficient is very large. For example, a chart shown by Bennett and Franklin (1954, p. 275) indicates that a correlation coefficient of +0.8 computed from a sample of 20 items would have a confidence belt extending from 0.6 to 0.9 for 95-percent probability. Because of this uncertainty, two correlation coefficients differing only by a few hundredths cannot be meaningfully compared. There also seems to be no justification for reporting correlation coefficients to more than two significant figures.

If the data can reasonably be assumed to be drawn from a normal bivariate distribution, then both correlation and regression analyses are appropriate. It is under this assumption that most of the examples in statistics texts are analyzed. However, regression is also appropriate under certain other conditions when correlation is not. The only assumptions required for regression are:

1. The deviations of the dependent variable about the regression line (for any fixed X) are normally distributed, and the same variance exists throughout the range of definition.
2. Values of the independent variable are known without error. The dependent variable is considered as an observation on a random variable, and the independent variable as some known constant associated with this random variable.
3. Observed values of the dependent variable are uncorrelated random events.
4. Each of the variables is homogeneous; that is, all individual values of a variable measure the same thing. Data are considered homogeneous if any subgroup to which certain of these data may be logically assigned has the same expected mean and variance as any other subgroup of the population. Neither variable need have a probability distribution in regression (but, of course, Y values corresponding to a fixed X are assumed to be normally distributed).

The end products of a regression analysis are two equations, Y = f(X) and X = f(Y) (usually only one is computed), because regression is directional. In contrast, correlation gives one index of the relation between variables.

The regression equation gives the average amount of change in the dependent variable corresponding to a unit change in the independent variable. Thus it gives more specific information than correlation. The regression coefficient can be tested to determine whether it is significantly different from zero, and this test is identical to the test of significance of the correlation coefficient (providing the data are drawn from a bivariate normal distribution).

4.4.1 Standard error

The reliability of a regression is measured by the standard error, which is the standard deviation of the distribution (assumed normal) of residuals about the regression line. (Figure 4.14 shows distribution of residuals.) By definition, the standard error is the same throughout the range of X. This standard error was called the standard error of estimate by Ezekiel (1950, p. 131). It is also referred to as the standard error of regression and as the standard deviation from regression.

The standard error of a prediction from regression is made up of three parts: the error of the mean, the error of the slope of the line, and the standard error of estimate. All three may be expressed in terms of the standard error of estimate so that the standard error of a prediction (S_p) is

$$S_p = S_e\sqrt{1 + \frac{1}{n} + \frac{(X - \bar{X})^2}{\Sigma(X - \bar{X})^2}}$$

where S_e is standard error of estimate, n is number of items in the sample, and X is the independent variable. Thus the error of a prediction increases with distance from the mean (Snedecor, 1948, p. 120)

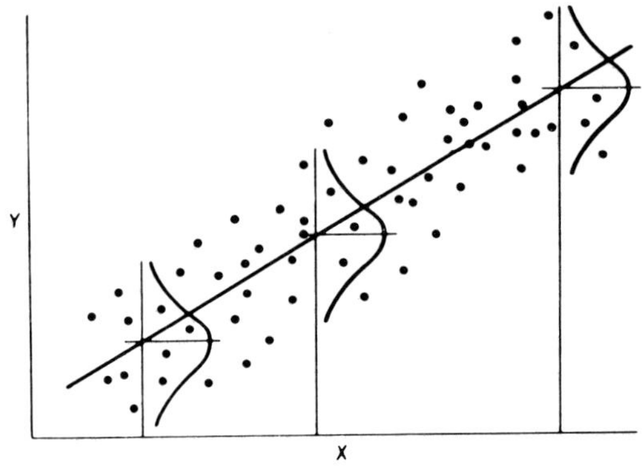

Fig. 4.14. Normal distribution of plotted points about the regression line.

4.4.2 Multiple correlation and regression

Most analyses require use of multiple correlation or regression. A multiple correlation is evaluated by partial correlation coefficients and by an index of total correlation. A partial correlation coefficient is an index of the degree of association between one independent variable and the dependent variable after the effects of the other independent variables have been removed.

In a multiple regression equation the regression coefficients are called partial regression coefficients. Each shows the effect on Y of a unit change in the particular independent variable, the effects of the other independent variables being held constant.

If the independent variables in a regression analysis are related to each other, the partial regression coefficients will be of a different magnitude from the simple regression coefficients. (The independent variables in a regression usually are related to each other as well as to the dependent variable.) See the section on "Application of the Regression Method" for elaboration on this subject.

The assumptions required for correlation are infrequently met in engineering problems and not generally met in hydrologic problems. Many of these problems to which the correlation method does not apply can be handled by the regression method because of the less restrictive assumptions. Thus the regression method may be used for such relations as that of concrete strength to time of setting, where neither value is randomly selected and neither variable has a probability distribution. Obviously the definition of such a relation is limited to the range of the data selected.

Under the above conditions the correlation coefficient does not apply but, of course, can be computed from the relation

$$r = \sqrt{1 - (S_e/S_y)^2}$$

where r = correlation coefficient, S_e = standard error of estimate, and S_y = standard deviation of the values of the dependent variable. From the above formula it can be seen that r depends on S_y, which depends on the range of data selected for problems such as the concrete strength relation to time of setting. Therefore, if the variables used in a regression are not randomly sampled, the computed value of r changes with the range of the arbitrarily selected sample and is therefore meaningless. Empirical verification of this statement is given by the data plotted in Figure 4.15. (These data were selected to demonstrate

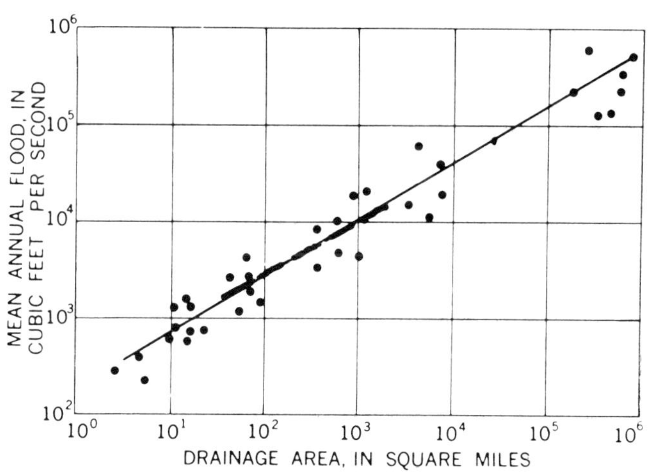

Fig. 4.15. Plot used to show the effect of sample range on computed correlation coefficient. Dashed line is the relation for 14 drainage areas ranging from 40 to 2,000 square miles.

this principle; the relation is not hydrologically significant.) Using all the points, the relation is computed to be

$$\log \text{MAF} = 2.27 + 0.59 \log \text{DA}$$

The standard error is 0.22 log unit and the computed correlation coefficient is 0.97. MAF is mean annual flood and DA is drainage area. If only the 14 points for drainage areas ranging from 40 to 2,000 square miles are used, the relation is

$$\log \text{MAF} = 2.31 + 0.57 \log \text{DA}.$$

This relation has a standard error of 0.23 log unit (almost the same as the previous standard error), but the computed correlation coefficient is 0.83, much lower than that obtained by using samples from a greater range. Obviously such variability in the correlation coefficient would render it unsuitable as a measure of the degree of relation for this type of application.

Regression is emphasized over correlation, not only because correlation is commonly inapplicable to particular hydrologic data but because regression provides quantitative answers to specific problems. In general, regression is preferred over correlation for hydrologic problems even when the data are suitable for a correlation analysis. Uses of regression analysis are:

1. To estimate individual values of the dependent variable corresponding to selected values of the independent variables.
2. To determine the amount of change in the dependent variable associated with a unit change in an independent variable.
3. To determine whether certain variables (which do not have probability distributions) are related to a dependent variable.

Correlation is most useful in theoretical studies and in time-series analysis.

4.4.3 Serial correlation

It has been pointed out that for a probability distribution to be valid the individuals must occur randomly. Hydrologic data such as daily stream discharges form a time series, that is, a sequence of values arranged in order of occurrence. The characteristics and analysis of hydrologic time series have been described by Dawdy and Matalas (1964). A common characteristic of a time series is the existence of a nonrandom element which produces a dependence between observations k units apart. This dependence is called serial correlation, and its degree is measured by the serial correlation coefficient.

First-order serial correlation is the dependence between observations adjacent in time; the kth order is the dependence between observations k units apart. A plot of the serial correlation coefficient against order is a correlogram (Dawdy and Matalas, 1964).

To determine the serial correlation, the time series is related to itself offset k units. For example, the time series in the first column below is related to itself shifted one observation to obtain the first-order serial correlation coefficient.

x_1 —

x_2 x_1

x_3 x_2

. .

. .

. .

x_n x_{n-1}

Computational details are the same as for the relation between two variables and are given in the section 4.5.3. A test of significance of a first-order serial correlation coefficient is given by Dawdy and Matalas (1964).

4.5 REGRESSION METHODS

The previous section described regression in general terms and concluded with some uses of regression. This section describes the computation and interpretation of regression equations, both analytical and graphical, and some characteristics of the regression method.

4.5.1 Regression models

We begin a regression problem with a dependent variable which we want to predict from one or more independent variables. The independent variables are values or characteristics which seem to be physically related to the dependent variable. Next we need a model which describes the way in which the independent variables are related to the dependent variable. The model should be in accord with known physical principles, but its exact form may be dictated by the data used.

Using a dependent variable, Y, and independent variables, X and Z, the equations and graphs of some more common regression models are shown in Figure 4.16. Joint relations, those which include a variable which is the product of two other variables, have been discussed in detail by Ezekiel and Fox (1959). The product of two variables is called an interaction term. Combinations of the models shown in Figure 4.16 can be used to describe more complicated relations, and the equations can be readily extended to include additional independent variables.

Having selected a suitable model, the coefficients in that model equation are computed from sample data by the method of least squares as described subsequently.

Note that although two of the graphs in Figure 4.16 are curved, all of the model equations are in linear form. This linearity of the model equation is a requirement for direct least-squares solution. Linearity can sometimes be attained by transforming the variables.

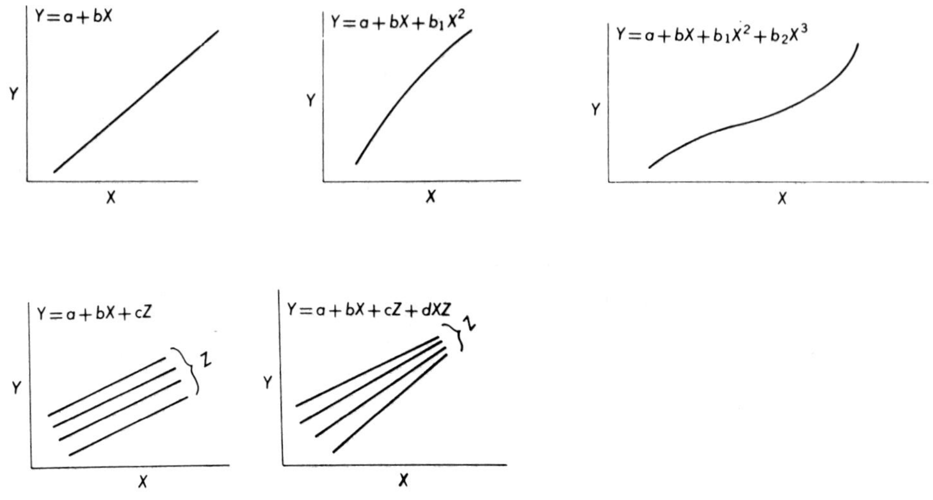

Fig. 4.16. Equations and graphs of some common regression models.

4.5.2 Transformations

There are two principal reasons for transforming data before analysis: (1) to obtain a linear regression model, and (2) to achieve equal variance about the regression line throughout the range.

We have seen from Figure 4.16 that the equations for certain two-variable regressions may be linearized without transforming the variables. The method is known as polynomial regression in which additional variables in successively higher powers of the independent variable are added to the model. But suppose we know or postulate that a relation should be of the form

$$Y = aX^b$$

By taking logarithms of both sides of the equation the resulting linear equation is obtained:

$$\log Y = \log a + b \log X$$

in which log a and b are constants which can be computed by a least-squares analysis using the variables log Y and log X. Likewise the relation

$$Y = ab^X$$

can be transformed to

log Y = log a + X log b

where log a and log b are the constants and log Y and X are the variables. Other transformations are sometimes used, but the logarithmic transformation is by far the most common.

The second reason for transforming data, and the more important one, is to achieve equal variance about the regression line. One of the assumptions basic to the regression method is that the distribution of errors about the regression line is normal and constant throughout the range (Fig. 4-14). Again a log transformation is often used. For example, the graph on the left of Figure 4.17 shows increasing scatter of points with increasing rainfall. But when the variables are plotted on the log chart (right graph), the scatter of points is almost uniform throughout the range. Thus, if it had been desired to carry the analysis beyond the graphical presentation, a transformation of the variables should have been made. (Ordinarily the variables would be reversed on the chart if a regression were to be made because runoff is the dependent variable.)

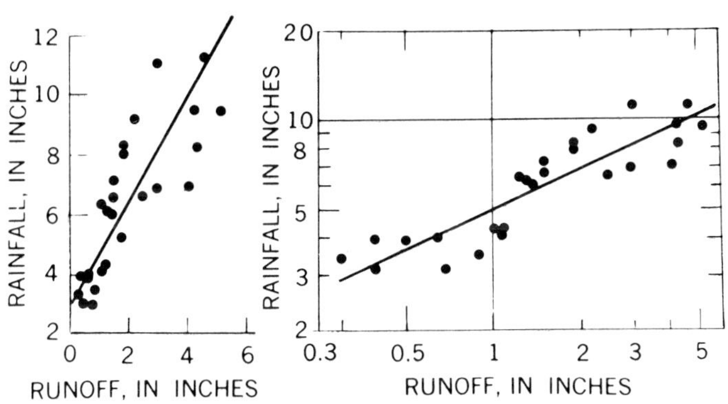

Fig. 4.17. Data plotted on natural and log scales showing the achievement of equal variance about the regression line by use of the log transformation.

Other reasons for transforming data are discussed by Acton (1959, p. 219-223).

4.5.3 Example of simple linear regression

Computation of a regression equation by the method of least squares using the model Y = a + bX is demonstrated using the data given in Table 4.6. That table also shows computations of means, cross products, and squares. The individual cross products and squares need not be recorded; the sum of cross products, or squares, can be cumulated on a calculator. The coefficients a and b in the regression equation, and the standard error of estimate are computed as shown below.

$$b = \frac{\Sigma XY - \frac{\Sigma X \Sigma Y}{N}}{\Sigma X^2 - \frac{(\Sigma X)^2}{N}}$$

$$b = \frac{192{,}042 - \frac{(1{,}801)(1{,}799)}{18}}{189{,}291 - \frac{(1{,}801)^2}{18}} = 1.325$$

$$a = Y - bX = 99.94 - (1.325)(100.06) = -32.6$$

Then the equation of the least-squares line is

$$Y = a + bX = -32.6 + 1.32X$$

$$\text{Variance of X} = S_x^2 = \frac{\Sigma X^2 - \frac{(\Sigma X)^2}{N}}{N-1} = \frac{189{,}291 - \frac{(1{,}801)^2}{18}}{17} = 534.76$$

$$\text{Variance of Y} = S_y^2 = \frac{\Sigma Y^2 - \frac{(\Sigma Y)^2}{N}}{N-1} = \frac{197{,}373 - \frac{(1{,}799)^2}{18}}{17} = 1{,}033.71$$

$$\text{Error Variance} = S_{y \cdot x}^2 = \frac{N-1}{N-2}[S_y^2 - b^2 S_x^2] = \frac{17}{16}[1{,}033.71 - (1.325)^2(534.76)] = 100.8$$

$$\text{Standard error of estimate of Y} = S_{y.x} = 10.0$$

$$\text{Correlation coefficient} = r = \frac{bS_x}{S_y} = (1.325)\ \frac{23.13}{32.15} = 0.95.$$

The locus of the regression equation and the data used are plotted in Figure 4.18.

(i) Test for significance. Although it is apparent from the plot of Figure 4.18 that there is a significant relation between the variables, this is not always the case. Sometimes it is desirable to test for significance. The regression coefficient, b, is the slope of the line. If that slope is zero the regression is meaningless. We start with the hypothesis that the true value of b is zero and assume that the distribution of sample regression coefficients is approximately normal with a zero mean and a standard deviation, S_b, computed as

$$S_b^2 = \frac{S_{y.x}^2}{\Sigma X^2 - (\Sigma X)^2/N} = \frac{100.8}{189{,}291 - (1{,}801)^2/18} = 0.011$$

and $S_b = 0.105$

TABLE 4.6

Data and computations for example of two-variable regression

	Percent of Mean				
Year	Runoff (Y)	Precipitation (X)	XY	X^2	Y^2
1928	125	110			
9	67	73			
30	68	74			
1	71	91			
2	118	108			
3	144	130			
4	169	152			
35	138	134			
6	102	98			
7	91	90			
8	125	119			
9	87	77			
40	84	100			
1	58	84			
2	79	85			
3	124	115			
4	62	70			
1945	87	91			
Totals	1799	1801	192,042	189,291	197,373
Mean	99.94	100.06			

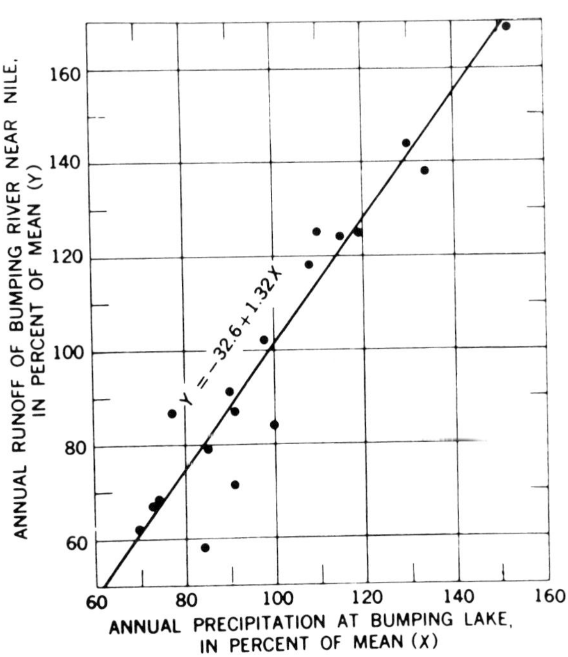

Fig. 4.18. Plot of data from Table 4-6 showing computed regression line.

The assumed distribution of regression coefficients is in Figure 4.19. If we sample from this normal population, we would expect about 2/3 of the samples to fall within plus or minus one standard deviation of the mean, and about 95 percent of the samples to fall within plus or minus two standard deviations of the mean. Since we have only one sample, we can say that (1) if it falls within two standard deviations of zero, we have no reason to believe that the true value is not zero, and (2) if it falls farther than two standard deviations from zero, it is unlikely that the mean is zero. The second condition would lead to

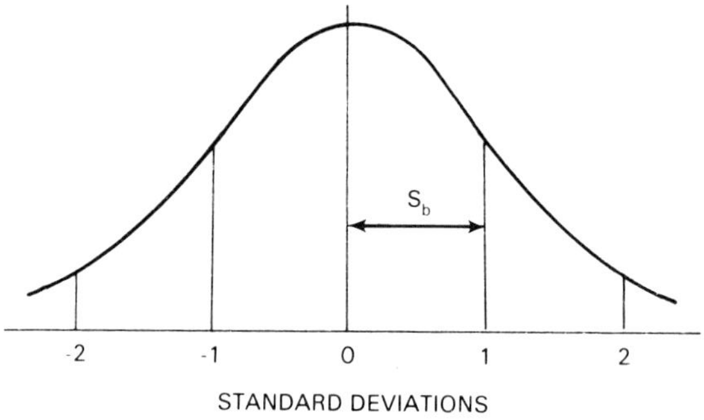

Fig. 4.19. Distribution of regression coefficients.

rejection of the hypothesis at the 95 percent probability level; that is, we conclude that the regression coefficient is significantly different from zero, or as commonly stated, significant.

In the above example b is 12.6 standard deviations from the mean (1.325/0.105) and thus is highly significant. For a normal distribution 95 percent of the samples would lie within 1.96 standard deviations plus and minus of the mean. The applicable distribution for small samples is the t distribution in which the probability associated with distance from the mean to any given multiple of the standard deviation changes with the sample size as measured by the degrees of freedom. For the above example there are 18−2 = 16 degrees of freedom for which t = 2.12 standard deviations rather than the 1.96 for the normal distribution; thus the regression coefficient is significant at the 95 percent level if it is more than 2.12 standard deviations from the mean of zero.

Most regression analyses are made by digital computer. The output includes tests of significance of the regression coefficients.

(ii) Degrees of Freedom. In the above example the degrees of freedom were two less than the number of items. One illustration of the concept of degrees of freedom can be explained from regression. The purpose of regression is to predict, not to indicate what happened in the past. If a curve is drawn through all points used in its definition, no check on the position of the curve is provided; that is, there are no degrees of freedom. A linear relation between two variables requires two points to define it; therefore, the number of degrees of freedom is the number of items used in its definition minus 2. The number of lost degrees of freedom in any regression is equal to the number of coefficients in the regression equation.

4.5.4 Multiple linear regression

The relation of a variable to two or more other variables can be obtained by multiple regression using a linear model. The least-squares method is employed; it provides the equation for which the sum of squares of deviations of the data from the values given by the equation is a minimum. Details of computation methods are given in statistics texts; the required computations are extensive. Consequently most multiple regression equations are obtained by digital computer. The output of a regression program includes the coefficients in the equation, the standard error of estimate of the dependent variable, tests of significance of the coefficients, and optionally the residuals from the equation. Residuals are the differences between the data and the corresponding values given by the equation.

Some computer programs will select, from the independent variables entered, those which are most highly related to the dependent variable. This may be done in two ways. The step-backward program first defines the equation using all the variables, then defines other equations by successively deleting the least significant variable in the preceding equation. A typical output of the step-backward method using log-transformed data is:

$$\log Q = -0.841 + 1.01 \log A - 0.12 \log L + 0.17 \log E + 0.38 \log P + 0.80 \log I$$

$$\log Q = -0.569 + 1.01 \log A + 0.12 \log L + 0.23 \log E + 1.36 \log I$$

$$\log Q = -0.538 + 1.01 \log A + 0.25 \log E + 1.27 \log I$$

$$\log Q = 0.086 + 1.01 \log A + 0.10 \log E$$

Not all the variables in these equations are necessarily statistically significant. Those containing nonsignificant variables should not be used.

The step-forward program provides one equation containing only those independent variables that are statistically significant at the prescribed probability level.

In evaluating regression equations with different numbers of independent variables, a distinction should be made between statistical significance and practical significance. An independent variable may be statistically significant at the 95 percent or 99 percent level and yet have a negligible effect on the computed values of the dependent variable. such an independent variable ordinarily would not be included in the final equation.

Selection of a regression equation from several which differ only in the number of independent variables can be made on basis of the sizes of the standard errors. A slight reduction in standard error because of including an additional variable does not necessarily indicate that the equation with the smaller standard error will produce more reliable estimates.

Further discussion of multiple regression is given in section 4.5.8.

4.5.5 Graphical regression

The assumptions required of graphical regression are the same as those required for analytical regression. The results of a graphical regression can be expressed mathematically if no restrictions are added to the graphical analysis, and the standard error can be estimated.

Graphical regression is less restrictive than analytical regresion in that the model need not be completely specified in advance. In fact, if an analytical model cannot be selected on a physical basis, it is conventional to prepare a preliminary graphical regression which will indicate an appropriate model. For example, consider the four data plots of Figure 4.20; the upper left one indicates use of the model

$$Y = a + bX$$

The upper right requires

$$Y = a + bX + b_1X^2$$

where the direction of curvature determines the sign of b_1. The lower left indicates the need for a transformation unless the divergence can be explained by an additional variable. The lower right shows no relation between Y and X, and, if only a two-variable relation is being considered, no further analysis would be made. A relation, however, between Y and X in the fourth plot may be obscured by the effect of another variable Z which has not been included. This aspect is discussed under Graphical Multiple Regression.

The preparation of simple linear relations between two variables is well known but the regression line is not necessarily the same line as one would draw through the plotted points. There are two regression lines, one for $Y = f(X)$, and another for $X = f(Y)$ (Fig. 4.21). The structural line, which balances the plotted points in both directions, has a slope approximately midway between the two regression lines. The differences in slope among the three lines depend on the degree of correlation of the variables. For perfect correlation all three lines have the same slope. Regardless of the correlation, both regression lines pass through the mean; the structural line may or may not pass through the mean.

To approximate the regression $Y = f(X)$, (1) group the points by small increments of X, (2) estimate the mean of each group in the Y direction, and (3) draw a line which averages these means. The procedure can be understood by referring to Figure 4.21 and remembering that the distribution of points about the regression line in the Y direction is assumed to be the same throughout the range.

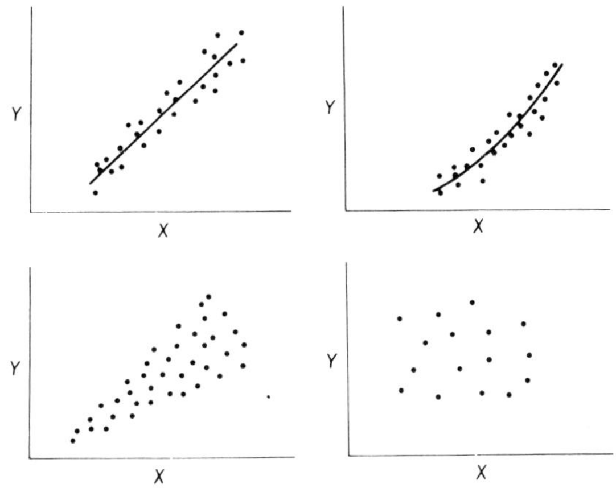

Fig. 4.20. Four possible outcomes of plotting Y against X.

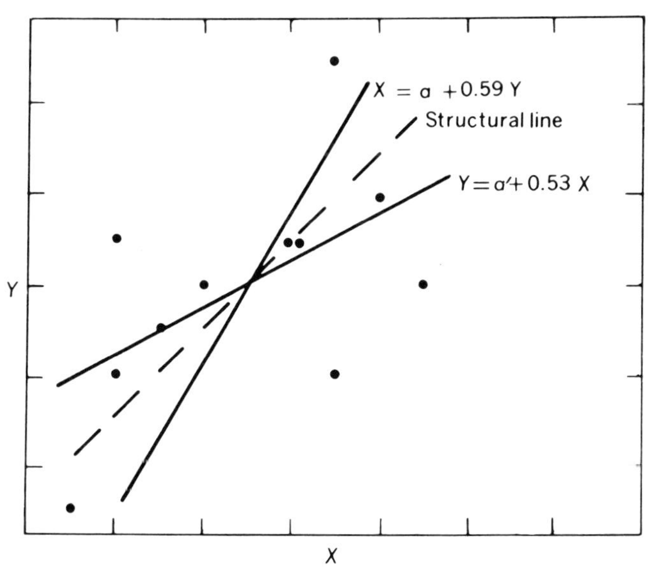

Fig. 4.21. Plot showing the two regression lines and the structural line.

Obviously that assumption cannot be true for a small number of points, but it is the condition which we try to approximate. The regression line of Y = f(X) will have a flatter slope than that of a line drawn to balance the points in both Y and X directions.

The standard error of estimate of a graphical regression can be estimated readily. Remembering (1) that the standard error of estimate is the standard deviation of plotted points about the regression line, (2) that two-thirds of the points should be within one standard deviation on each side of the mean of a normal distribution, and (3) that a regression line theoretically passes through the mean value of Y corresponding to any value of X, then two lines, parallel to the regression line and one standard deviation above and below (in the Y direction), should encompass two-thirds of the plotted points. In practice it is simpler to draw the lines so as to exclude one-sixth of the points above and below, and then use the average of these two deviations from the mean as the estimated standard error of estimate. The procedure is illustrated in Figure 4.22 for a log relation. The standard error can be described in log units but more commonly is expressed in percent. This value is readily obtained by using dividers to lay off one standard error above and below a cycle separation if the relation is plotted on log paper. The percentages are measured as shown in Figure 4.22. The standard errors in percent can be computed by $100\ (10^S-1)$ and $-100\ (1-10^{-S})$ where S is the standard error in log units.

For regressions on arithmetic plots, the standard error will be in the same units as Y and can be read from the plot.

The reliability of the graphically determined standard error is influenced by two factors having opposite effects. If the graphical-regression line has a steeper slope than the least-squares regression line, the graphical standard error will be larger than the computed standard error. If we assume that the graphical line of relation is the same as the least-squares line, the graphically computed standard error will underestimate the computed standard error when a few plotted points are far from the line but the majority are close. In any case the graphically determined standard error is only an approximation but is adequate for many problems.

4.5.6 Graphical multiple regression

There are two general methods of graphical multiple regression. The method of deviations is based on a model of the type

$$Y = a + b_1X_1 + b_2X_2 + \ .\ .\ .\ b_nX_n,$$

or a similar one allowing for curvilinearity. This method is probably the simplest and most useful one available.

The coaxial method of graphical multiple regression formerly used for runoff-precipitation relations, is a more flexible method than the method of deviations in that it allows both for interactions and curvilinearity. However, these advantages are obtained at the expense of much additional work and at the loss of a simple method of evaluating the reliability of the result. Linsley, Kohler

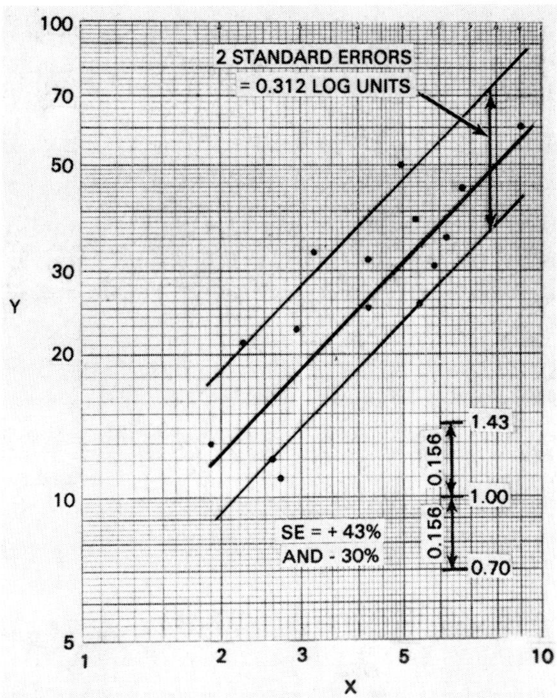

Fig. 4.22. Method of estimating the standard error of a graphical regression.

and Paulhus (1949, p. 650-655) described the procedure in detail. The descriptions of graphical multiple regression in this section refer to the method of deviations.

The purpose of multiple regression is to determine how a dependent variable changes with changes in two or more independent variables. This problem cannot be solved by considering one independent variable at a time if the independent variables are correlated with each other. This statement can be verified by analyzing the following synthetic data:

No.	Y	X_1	X_2
1	500	100	25
2	250	150	160
3	300	50	30
4	100	30	110
5	200	100	150
6	200	20	20
7	50	50	700

Assume that the logarithms of the variables are linearly related. This calls for plotting on log paper. First make a graphic comparison between Y and X_1 by plotting the appropriate data (Fig. 4.23). (In statistical work the dependent variable is usually plotted on the ordinate scale.) Also plot Y against X_2. These plots indicate that Y cannot be estimated reliably from either parameter.

Now determine the relation between Y and both of the other variables. The procedure is as follows:

1. On Figure 4.23, plot 1, write beside each point the corresponding value of X_2. It will be seen that the high values of X_2 tend to be on one side of the group and the low values on the other. This is an indication that X_2 values are related to Y. Draw a straight line through the points in such a way that it represents roughly some constant value of X_2. The line probably will not balance the plotted points.
2. Plot deviations (also called residuals) of Y from the straight line of plot 1 against X_2 as the abscissa on plot 3. The deviations may be scaled from plot 1 or transferred by dividers. Because they are ratios, they should be measured above or below 1.00 on plot 3.
3. Draw a straight line averaging the points on plot 3.
4. Measure the deviations of the points from the curve of plot 3 and replot them on plot 1. These deviations are measured from the straight line in plot 1 and define the relation between Y and X_1 with the effect of X_2 removed. Sometimes these replotted points are not randomly distributed about the line, in which case the line should be redrawn and the whole process repeated (not necessary in this example). When a satisfactory balance is attained the regression is complete. The scatter of the adjusted points about the line of plot 1, is a measure of the error. The standard error of a graphical multiple regression may be approximated by using the adjusted points, as described in section 4.5.5. The line on plot 1 is the relation between Y and X_1 for the X_2 value at which the line of plot 3 crosses the 1.0 line (X_2 = 66). The relation of Y to X_1 for any other value of X_2 will be a line parallel to the line of plot 1, at a position defined by the curve of plot 3 for the desired value of X_2.

The example used gave much better results than ordinarily would be expected in hydrologic analyses. The data were manufactured (1) to illustrate the procedure and (2) to point out that a good relation may not be recognized if only two variables at a time are studied.

Graphical regressions involving more than two independent variables can be made. The residuals from each line are plotted against the next variable until all variables are used. Then the residuals from the last line are replotted

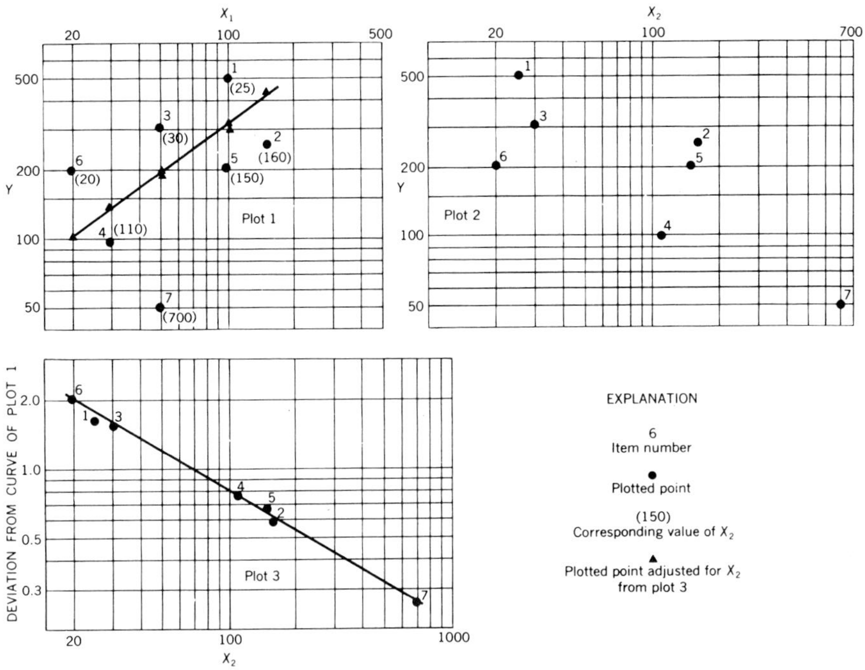

Fig. 4.23. Example of graphical multiple regression.

from the first as described in step 4. In practical work it is usually difficult to define the effects of more than three independent variables, particularly when the influences of one or two of the variables are small.

Linear regression should be used whenever the plotted points do not definitely define a curve and when no physical reason is known for expecting the relation to be curved. If a curve or curves are indicated by both of the above criteria, then curves should be used. Complicated curves require four or more points for definition. They should be avoided when only relatively few points are available to define the relation.

Graphical multiple regressions need not be made on logarithmic paper. Arithmetic and semilog plots can be handled as readily. Figure 4.24 relates summer runoff to spring water content of the snowpack and to summer precipitation. The graphical procedure is the same as in the first example, except that deviations are measured in the same units as the dependent variable and the deviation scale must have its center at zero with positive values above and negative values below. Obviously the mathematical model describing this relation is different from one for a graphical relation developed on log paper.

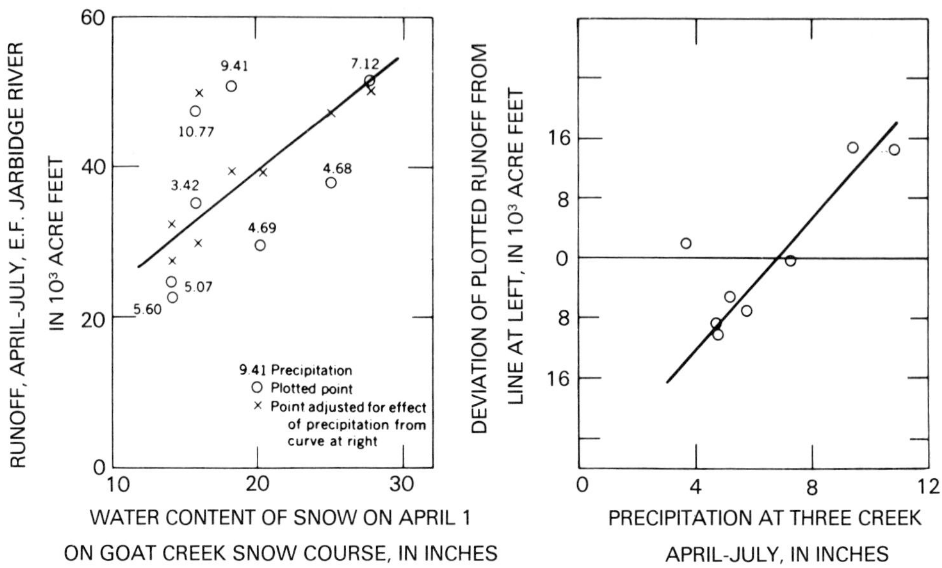

Fig. 4.24. Example of graphical multiple regression using arithmetic scales.

The plotting paper selected for a particular problem should be that on which the distribution of the dependent variable for a fixed value of the independent variable is approximately the same for all values of the independent variable as in Figure 4.14. This criterion is more important than that of attaining linearity.

4.5.7 Graphical vs. analytical method

A graphical method is particularly useful for exploratory work and for making preliminary estimates. The graphical method has the following advantages:

1. It is rapid.
2. It helps define the appropriate model.
3. It points out the need for transformations, if any.
4. It brings attention to extremely wild points if they exist in the data. (See the wild point in Fig. 4.24 which was given no weight in the analysis).

Disadvantages of a graphical method are:

1. Small effects of independent variables cannot be identified.
2. The number of independent variables is limited to about three because of the cumulative effect of inaccuracies in plotting and in locating the lines.
3. Tests of significance of the effects of individual variables are not available.
4. The resulting relation involving three or more variables is confusing to the user unless expressed mathematically or replotted in another form.

An analytical method has the following advantages:

1. For the model used, it gives the best estimate of the equation constants, and of the standard error.
2. It allows testing of the coefficients for significance.
3. Results can be presented in a clear, concise manner which most hydrologists can understand.
4. Results are unique for the model and sample used; different investigators would get the same results.

Disadvantages of an analytical method are:

1. Computation is time-consuming unless the analyst has access to a digital computer and knows how to enter his data.
2. The existence of wild points or of nonhomogeneous data is not readily noticed.
3. The selected model may not be appropriate.
4. No procedure is available for giving a point less than full weight.

In general a graphical method is useful for examining the data and defining a suitable model; final conclusions are usually based on a computed relation.

4.5.8 Application of the regression method

An analytical problem to be solved by regression involves (1) selection of factors which are expected to influence the dependent variable, (2) describing these factors quantitatively, (3) selecting the regression model, (4) computing the regression equation, the standard error of estimate, and the significance of the regression coefficients, and (5) evaluating the results.

Selection of the appropriate factors should not be a statistical problem, but statistical concepts must enter into the process. If the analyst merely wants to know the relation of annual precipitation to annual runoff, he can proceed directly to selection of a model. But if his problem is to make the best possible estimate of runoff, he will include other factors, some of which may be related to each other as well as to runoff. The problem of determining if certain factors are related to the dependent variable requires careful selection of indices describing these factors quantitatively. These indices should accurately reflect the effects, and no two should describe the same thing. It is a characteristic of regression that if a factor is related to a dependent variable and this factor is entered in the regression model twice (as two different variables), the effect on the dependent variable will be divided equally between the two. Thus, if the total effect is small, the result of dividing it in two parts may be to produce nonsignificance in each of the parts. Likewise, several closely related variables may compute as nonsignificant, whereas one properly selected index would show a real effect. Thus, the independent variables should be selected with considerable care; the shotgun approach should not be used.

Another consideration in selection of variables is to avoid having a variable, or a part thereof, on both sides of the equation. Such a condition may be acceptable for certain problems, but the results must be evaluated carefully. A spurious relation may result, or the relation may be correct but its reliability difficult to assess. Benson (1965) described ways in which spurious relations may be built into a regression.

The user of the regression method should understand the effect of related independent variables on the computed regression coefficients. If the independent variables are entirely unrelated, the simple regression coefficients and the corresponding partial regression coefficients would be the same. However, such conditions rarely occur in nature. The multiple regression method provides a way of separating the total effect of the independent variables into the effect of each independent variable and an unexplained effect. Consider the simple regression

$$Y = a + b_1X_1 + \text{error}, \tag{1}$$

where Y also is affected by another variable, X_2 , which is related to X_1 . The regression using X_1 and X_2 will be

$$Y = a' + b_1'X_1 + b_2X_2 + \text{error}, \tag{2}$$

where $b_1' \neq b_1$. If X_1 and X_2 are the only variables affecting Y (and the effects are linear), then equation 2 completely describes Y, and b_1' and b_2 are the true values of the regression coefficients (except for sampling errors). If X_1 and X_2 are positively correlated with each other and with Y, consider the effect on the magnitude of b_1. For each value of X_1 in equation 1, Y will appear to be more closely related than it actually is because X_2 increases with X_1 and its influence on Y is real though unmeasured. Therefore the regression coefficient b_1 is larger than its true value b_1'.

Similar changes in b_1 and b_2 would occur if another factor, related to X_1 and X_2 and Y, were included in the regression. These changes in the magnitudes of the regression coefficients due to addition or deletion of a variable are characteristic of regression. They are sometimes interpreted as indicating that partial regression coefficients have no physical meaning. Such interpretations are not necessarily correct. If the variables used in the regression are selected on physical principles and the effects of each of the variables is appreciable, then the partial regression coefficients should be in accord with physical principles. In fact, it is good practice to compare the sign and the general magnitude of each partial regression coefficient with that expected. Benson (1962a, p. 52-55) made a thorough comparison of this kind.

The regression coefficients of certain variables may change sign when another related variable is added to or deleted from the regression. This effect may result because (1) the subject variable is not a good index of the physical feature represented, (2) the effect of the variable is small relative to the sampling error, (3) the variable is so highly correlated with one or more other variables in the regression that its real effect is obscured, or (4) the range of the variable sampled may be too small to define a significant effect.

A regression equation does not imply a cause-and-effect relation between the independent variables and the dependent variable. Both may be influenced by some other factor not readily measured. However, there should be some physical tie between the variables if the results can be considered meaningful.

Selection of a regression model usually begins with a graphical analysis. A model which plots as a straight line is commonly used unless there is strong evidence to the contrary.

If the sample data exist near an asymptote or near a maximum or minimum point on the curve, a simple model may be inadequate to describe the relation and a more sophisticated one may not be justified unless many data are available. An example showing the characteristics of three common models when applied to data defined near zero is given in Figure 4.25. Physical considerations suggest that neither b nor Q_7 should be less than zero and that the line should be curved. The zero limitation can be obtained by using the variables log b and log Q_7 and thus making the curve asymptotic to zero on both axes. The addition of a term $(\log Q_7)^2$ will provide the necessary curvature. The regression equation using these three variables is the top one on Figure 4.25. It is not a good fit to the data.

Next, assume that it is not necessary that the curve by asymptotic to $Q_7 = 0$. Then a semilog model using the variables log b, Q_7, and Q_7^2 would be appropriate. But the equation based on this model reaches a maximum too soon and is a very poor fit. As a last resort assume a simple model with the variables b, Q_7, and Q_7^2. This equation is a good fit to the data.

The mechanics of computing the regression equation, the standard error, and the tests of significance have been described. One important task remains, that of evaluating the results. First, the analyst should recognize that the regression equation developed, even though it is a good fit to the data, is not necessarily correct if extrapolated. For example, the curve corresponding to the bottom equation of Figure 4.25 is a good fit to the seven points but increases directly with Q_7 for values of Q_7 greater than 7. On the other hand, the dashed curve fits the lower five points but becomes asymptotic to zero as Q_7 increases. Available information does not indicate which extrapolation is more nearly correct.

The signs of all significant regression coefficients should be in accord with physical principles. The regression is not necessarily incorrect if they are

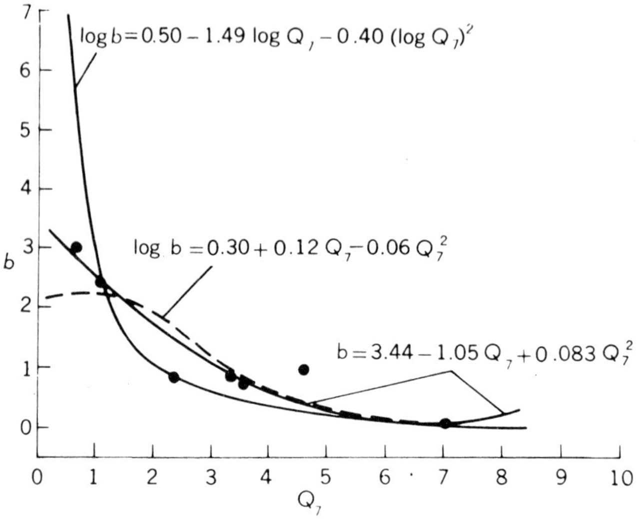

Fig. 4.25. Equations and graphs of 3 models based on the plotted data.

not; the nonconformity may be due to interrelations among the independent variables. Such a regression may be useful for estimating values of the dependent variables from known values of the independent variables, and the reliability of the results, if within the defined range of the regression, can be computed.

The more difficult problem of determining whether a particular variable is related to the dependent variable may not have a definite answer. Even though a regression coefficient is statistically significant, there is a small probability that this result occurred by chance. Other samples could produce conflicting results.

4.6 CHARACTERISTICS OF HYDROLOGIC DATA

Streamflow is a continuous process which varies with time, and thus streamflow data are said to form a time series. A plot of streamflow against time would show a pattern of variation recurring each year; that is, high flows tend to occur at particular times of the year and low flows at others in response to climatological characteristics which also vary seasonally.

Because streamflows are not discrete values, we need to chop the hydrograph into pieces which we will consider as individual streamflows. The particular pieces we use have certain characteristics which must be considered in analysis. The most common piece is the daily mean discharge. A daily mean discharge is related to the discharge of the previous day and lies within a range which depends on the time of year. In statistical terms, daily mean discharge is a serially correlated variable, that is, it is nonrandom. The daily mean discharges for a year are also not homogeneous; they are more likely to be larger at one time of the year than at another. Data are considered homogeneous if any subgroup to which certain of these data may be logically assigned has the same expected mean and variance as any other subgroup of the population.

Monthly mean discharges for different calendar months are also serially correlated and nonhomogeneous. Annual mean discharges may be homogeneous values. They may or may not be serially correlated, depending on the amount of basin storage at the time that the hydrologic year begins.

Instead of a streamflow variable made up of adjacent segments of a hydrograph, we may consider variables such as July mean, annual peak discharge, or annual minimum flow. These variables are made up of one individual from each year and thus are independent of the yearly cycle of streamflow. They are also independent of each other (with the possible exception of annual minimum flows which may include effluent from ground-water recharge of a previous year).

Precipitation, temperature, sediment discharge, water quality, transpiration, evaporation, and solar radiation vary throughout the year; indices describing them may be nonrandom and nonhomogeneous.

Obviously the distinction between random and nonrandom data and between homogeneous and nonhomogeneous data is not always clear. The analyst will have to determine whether the effects of possible moderate nonrandomness or nonhomogeneity will invalidate the conclusions of his particular analysis. It is important that the character of the data be considered in designing the analysis and in interpreting the results.

Another type of variable used extensively in hydrology cannot be considered to have a probability distribution, or even to be drawn from a population as thought of in the usual sense. Basin characteristics such as drainage area, slope, elevation, and vegetal index are in this category. (It is possible to conceive of certain physiographic parameters as random variables, but rarely can the available sample be considered randomly selected or representative.)

Time is sometimes used as a variable in regression. It has no distribution and is used only as a substitute for the real factor or factors (which are unknown or cannot be expressed by indices) associated with changes in a dependent variable.

4.6.1 Effects of data characteristics on analysis

We prepare a frequency distribution of daily mean discharges from several years of data; this is the duration curve. The individual values are nonrandom and nonhomogeneous. Therefore the duration curve cannot be considered a frequency curve. The probability of exceeding a certain value on a particular future day depends both on the preceding value and on the time of year. Thus, the duration curve is merely the distribution of daily means that has occurred. It can be considered an estimate of the distribution during a future period several years long.

On the other hand, frequency curves of annual flood peaks can be interpreted as probability curves because the individuals are unrelated and usually homogeneous (but see section 8.5). Most low-flow frequency curves can be similarly interpreted, but occasionally a serially correlated sample will be found and certain individuals may be nonhomogeneous (section 9.2).

The effect of using nonhomogeneous data in a regression problem is shown by Figure 4.26 in which is plotted 4 years of monthly mean discharge for each of

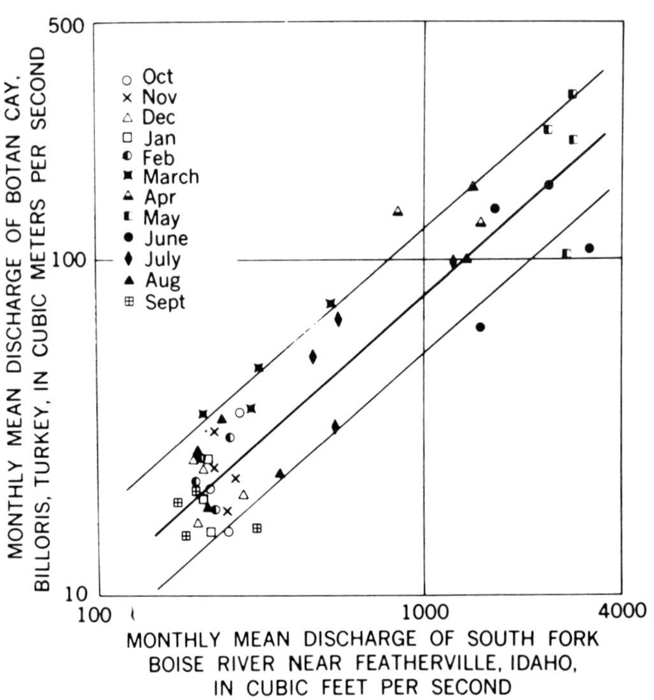

Fig. 4.26 Spurious relations from use of nonhomogeneous data.

the 12 calendar months for two stations, one in Turkey and one in Idaho, USA. The relation looks fairly good, but there is actually no relation between the two streams for a particular calendar month. The apparent relation using all calendar months arises because the yearly cycle of streamflow in Idaho resembles that in Turkey. Discharges in winter months are low and in spring snowmelt months are high.

Less extreme conditions are shown by relations between monthly mean discharges from contiguous basins. For example, there is no relation between monthly mean discharges of Lake Fork above Moon Lake, Utah, and Duchesne River at Provo River Trail, Utah, for January; there is a fair relation for June (Fig. 4.27). The data for all calendar months are obviously nonhomogeneous.

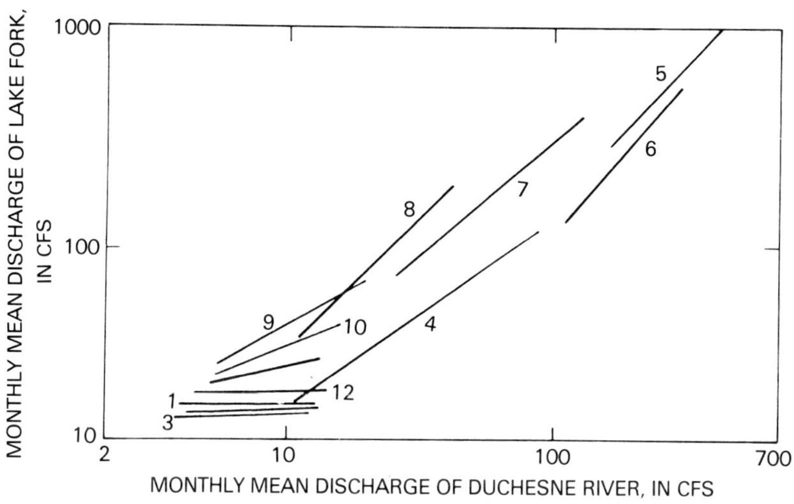

Fig. 4.27. Discharge relations for individual months, two Utah stations.

4.6.2 Outliers

Many factors influence the flow of a stream; some exert great influence at one time and none at another; most exert effects which are interrelated with effects of other factors. Only a few factors can be included in a regression used to estimate streamflow, and the effects of these factors are only approximated. Consequently there is a scatter of points about the regression line and occasionally a wild point occurs. Such wild points are called outliers in statistics, and statistical tests are available for use in determining whether or not a particular point should be rejected as not belonging to the group. It seems questionable whether outliers in hydrologic analyses should be rejected on the basis of a statistical test. Consider the wild point in Figure 4.24. If the precipitation had been about 7 inches instead of 3.4 inches, the point would not be wild. It is possible that precipitation at the higher elevations in the Jarbidge River basin was much greater than at Three Creek. If it were, the same thing could happen again and more weight should have been given to that point in the analysis. However, if some of the data for that year are found to be unreliable we could reject the point.

Acton (1959) devoted a short chapter to the rejection of unwanted data. He says, in part, "But the plain truth is that physical scientists and engineers need not be encouraged to ignore obstinate outlying data - rather they need to be held in check".

The term "outlier" as commonly used in hydrologic frequency analysis does not imply that the observation is in error or that it is from another population, only that it deviates widely from the others in the sample.

REFERENCES

Acton, F.S., 1959, Analysis of straight-line data: New York, John Wiley and Sons.

Bennett, C.A. and Franklin, N.L., 1954, Statistical methods in chemistry and the chemical industry: New York, John Wiley and Sons.

Benson, M.A., 1960, Characteristics of frequency curves based on a theoretical 1000-year record in Dalrymple, T., Flood-frequency analyses: U.S. Geol. Survey Water-Supply Paper 1543-A.

Benson, M.A., 1962a, Factors affecting the occurrence of floods in a humid region of diverse terrain: U.S. Geol. Survey Water-Supply Paper 1580-B, 64 p.

Benson, M.A., 1962b, Plotting positions and economics of engineering planning: Am. Soc. Civil Engineers Proc., Vol. 88, No. HY6, p. 56-71.

Benson, M.A., 1965, Spurious correlation in hydraulics and hydrology: Am. Soc. Civil Engineers Proc., V. 91, No. HY4, p. 35-42.

Beran, M.A., 1981, Recent advances in statistical flood estimation techniques in Flood Studies Report - Five Years On: Institution of Civil Engineers, London.

Chow, V.T., 1964, Frequency analysis, in Chow, V.T., Handbook of applied hydrology: New York, McGraw-Hill, p. 8-17.

Dalrymple, T., 1960, Flood-frequency analysis: U.S. Geol. Survey Water-Supply Paper 1543-A, 80 p.

Dawdy, D.R. and Matalas, N.C., 1964, Analysis of variance, covariance, and time series in Chow, V.T., Handbook of applied hydrology: New York, John Wiley and Sons.

Dixon, W.J. and Massey, F.J., 1957, Introduction to statistical analysis: New York, McGraw-Hill Book Co.

Ezekiel, M., 1950, Methods of correlation analysis, second edition: New York, John Wiley and Sons, Inc.

Ezekiel, M. and Fox, K.A., 1959, Methods of correlation and regression analysis: New York, John Wiley and Sons, Inc.

Gumbel, E.J., 1958, Statistics of extremes: New York, Columbia University Press, 375 p.

Kendall, M.G., 1952, The advanced theory of statistics: London, Charles Griffin and Co., Vol. 1, 457 p.

Langbein, W.B., 1960, Plotting positions in frequency analysis, in Dalrymple, T., Flood-frequency analysis: U.S. Geol. Survey Water-Supply Paper 1543-A, p. 48-51.

Lettenmaier, D.P. and Burges, S.J., 1982, Gumbel's extreme value I distribution: A new look: Journal of Hydraulics Division, ASCE, Vol. 108, No. HY4, April 1982, p. 502-514.

Linsley, R.K., Kohler, M.A., and Paulhus, J.L.H., 1949, Applied hydrology: New York, McGraw-Hill Book Co., Inc. 689 p.

McDonald, J.E., 1957, A critical evaluation of correlation methods in climatology and hydrology: Arizona Univ. Inst. of Atmospheric Physics Science Report 4, 35 p.

Riggs, H.C., 1961, Frequency of natural events: Am. Soc. Civil Engineers Proc., Vol. 87, No. HY1, p. 15-26.

Snedecor, G.W., 1948, Statistical methods, 4th edition: Iowa State College Press, 485 p.

Water Resources Council, 1982, Bull. 17B, Guidelines for determining flood-flow frequency: U.S. Water Resources Council, Washington, D.C.

Chapter 5

STREAMFLOW CHARACTERISTICS AT A GAGED SITE

5.1 INTRODUCTION

The continuously varying flow of a stream is characterized by the mean and variability of daily, monthly, and annual flows; by frequency characteristics of flood peaks, low flows, and other elements; and by base-flow recession curves.

5.2 MEANS

Streamflow records are published as daily mean flows from which the monthly means and the annual mean are computed. The streamflow record in Figure 3.16 (Chapter 3) shows daily means, monthly means, and the annual mean for water year 1979, and the average discharge for the period of record. Average discharge is the mean discharge computed from all complete years of record. The average of monthly mean discharges for a particular calendar month is the mean monthly (read as mean of the monthly) discharge. The mean monthly discharges for all twelve months show the flow distribution within the year. See Figure 5.1 for an example.

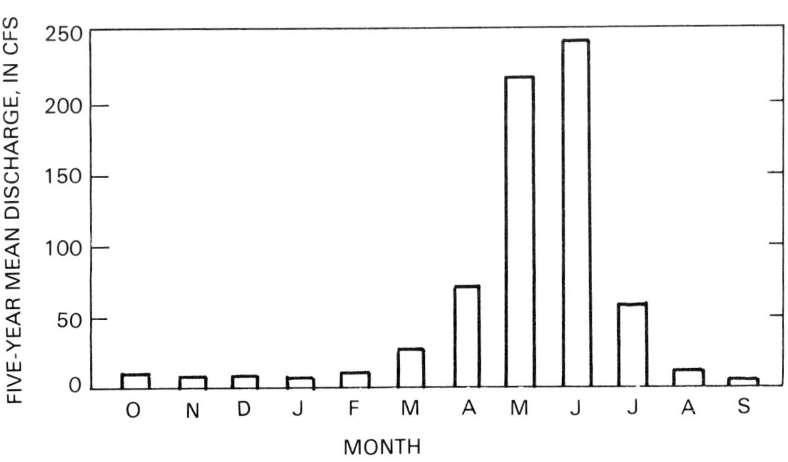

Fig. 5.1. Mean monthly discharges, E.F. Jarbidge River, Idaho.

5.3 FREQUENCY CHARACTERISTICS

Variability of flow can be described by the coefficient of variation or the index of variation, both of which can be computed readily; see Section 4.2.8. More commonly flow variability is described by a frequency curve from which selected points are used as descriptors. Furthermore, because most hydrologic data are asymmetrically distributed, the median flow (at 0.5 probability or 2-year recurrence interval) is often used as the index of central tendency rather than the mean. The frequency curve of annual mean discharges of Figure 5.2 shows the probabilities that an annual mean flow will be less than various curve values. For example, the probability is 10 percent that a future annual mean flow will be less than about 450 cfs.

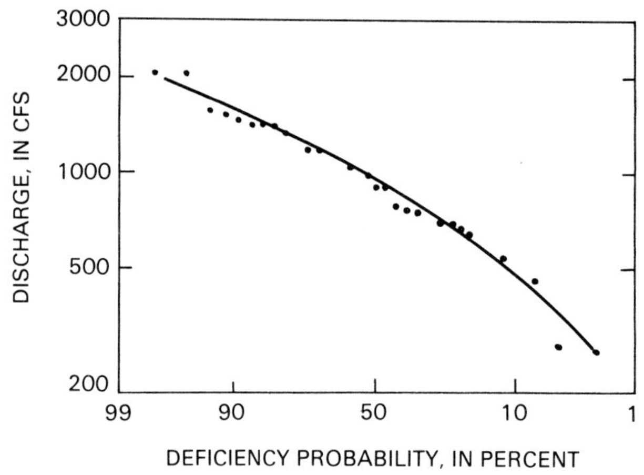

Fig. 5.2. Frequency curve of annual mean discharges for Red River, Tennessee.

Flood-frequency curves and low-flow frequency curves are usually defined in terms of recurrence interval; characteristics such as the 100-year flood and the 10-year low flow are common descriptors taken from these curves. Methods of preparing frequency curves are given in Chapter 4 and other examples are given in Chapters 8 and 9.

The reliability of a frequency curve depends on the number of years of data on which it is based, on the variability of streamflow, and on whether or not the flow regime was changed during or subsequent to the period of record. A requirement for validity of a frequency curve is that the data are all drawn from the same population; this is usually met by flows which are averaged over some period of time - such as monthly or annual means; it may not be met by flood peaks on ephemeral streams or by annual low flows in humid regions. A further requirement is that the data not be serially correlated. Thus an understanding of the basin hydrology is needed for proper interpretation of a frequency curve. The various factors that influence flood-frequency curves and low-flow frequency curves are discussed in the chapters on those subjects.

5.4 EXTENDING STREAMFLOW RECORDS IN TIME

A frequency curve based on 10 or 15 years of record may not be very reliable. Estimates for additional years can be made by regressions on precipitation or on a longer streamflow record concurrent with the one for the site to be estimated. These estimates are then combined with the flows of record to define another, and presumably better, frequency curve. Data in Table 5.1 are used to illustrate the procedure. Assume that annual flood peaks for Blackwater River are available only 1951-60. The regression relation between flood peaks for the two streams for that period is

$$\log B = 1.035 + 0.676 \log L$$

and the correlation coefficient is 0.85. This relation is used to estimate Blackwater flood peaks for 1922-50; the estimates and those for 1951-60 are shown in Table 5.1. Figure 5.3 shows frequency curves for Blackwater based on 1951-60; on the actual record; and on the combination of estimates for 1922-50 and actual for 1951-60. In this example, the frequency curve defined partly by estimates is poorer than the one based on only the 10 years of record.

TABLE 5.1

Annual floods, in cfs, on two Missouri streams.

Year	Lamine R.	Blackwater R.		Year	Lamine R.	Blackwater R.	
	Record	Record	From Regression		Record	Record	From Regression
				1941	18600	3890	8350
1922	15200		7280	2	21400	13400	9180
3	9300	9280	5230	3	60000	27900	18400
4	7640	10800	4580	4	32500	32400	12200
1925	10100	7060	5530	1945	12200	12600	6280
6	11300	10000	5960	6	14500	11300	7060
7	25000	17400	10200	7	14300	12900	6990
8	7620	21800	4570	8	32500	15600	12200
9	33000	54000	12300	9	12400	9760	6350
1930	7260	7990	4420	1950	12200	13200	6280
1	8500	3200	4920	1	65500	23900	19600
2	11200	9680	5930	2	9750	7100	5400
3	17800	6900	8110	3	5360	5880	3600
4	5190		3520	4	3830	3290	2870
1935	25000		10200	1955	8260	5170	4820
6	13200		6620	6	14000	3960	6890
7	22200		9410	7	6740	3150	4200
8	16600		7730	8	25500	8100	10300
9	40200	9810	14100	9	9980	3570	5480
1940	4280	5300	3090	1960	28700	16200	11200

Fiering (1963) has shown that unless the correlation coefficient is greater than about 0.8, use of regression estimates will not improve the estimate of variance. The example regression has an r of 0.85; and Figure 5.3 shows that the variance (slope of the line) is improved by use of estimates, but that the mean is not. Apparently the relation between flood peaks of the two streams during 1951-60 is considerably different than for other periods. In general, use of regression estimates produces an improved frequency curve only when the two variables are very highly correlated.

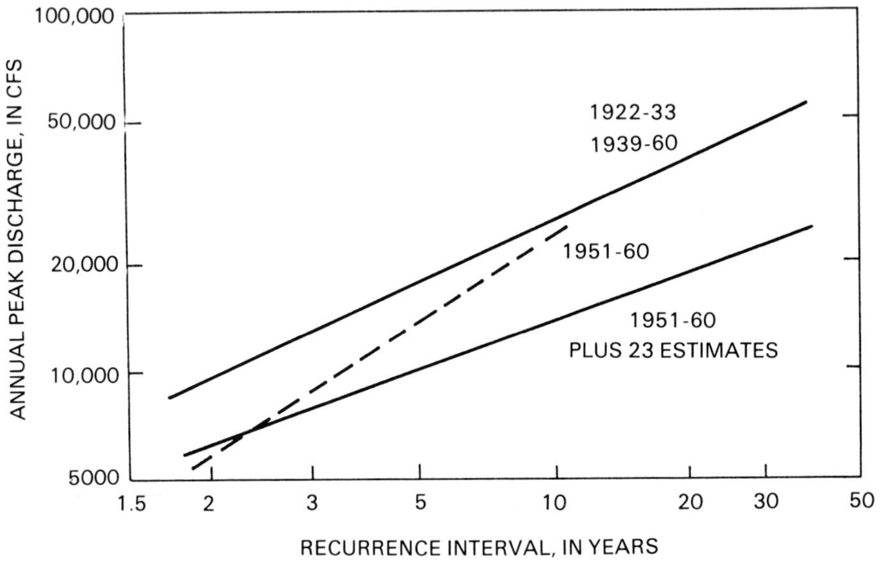

Fig. 5.3. Result of attempt to improve a flood-frequency curve by use of regression estimates.

Streamflow records may be extended in time for purposes other than frequency curve improvement. A mean for some specific past month or year may be needed. This can be obtained by simple regression or by regression models with both streamflow and precipitation as independent variables. Schneider (1961), Riggs (1964a) and Martin (1964) proposed and evaluated several models for estimating monthly mean discharges for months outside the period of record from discharge at a nearby gaged site and from the monthly precipitation totals on the two sites. Carroon (1970) extended monthly flow records at many sites in the upper Colorado River basin by a graphical method. The reliability of such a regression estimate is indicated by the standard error.

Rainfall-runoff models which account for the detailed disposition of rainfall may be used to extend streamflow records. These may be regression models in which runoff is related to storm magnitude, storm duration, and an index of antecedent soil moisture; or digital computer models which model the land phase of the hydrologic cycle. Both types of models require several years of concurrent streamflow and hourly precipitation records for calibration; their output is continuous daily streamflow, limited in time by the available precipitation record. Regression models are described by Moore (1968) and Sittner, et al (1969). There are many digital computer models for producing a continuous flow record from precipitation; most are some modification of the Stanford Watershed Model (Crawford and Linsley, 1966). Rainfall-runoff models produce synthetic streamflow records that are adequate for many purposes if the precipitation record is representative of precipitation on the basin. But the synthetic record is not reliable enough to improve frequency curves of daily extremes. A rainfall-runoff model that synthesizes storm hydrographs for the purpose of extending a flood record is described in section 8.7.

5.5 FLOW-DURATION CURVES

The flow-duration curve is a cumulative frequency curve based on daily discharges for one or many complete years of record. The usual graphical method of defining a frequency curve by plotting each point is not feasible when there are many points; in this application there are 365 for each year. Consequently, the numbers of days having discharges within each of 20 or 30 ranges of discharge are determined, usually by computer. These numbers are converted to percentages of the total, the percentages are cumulated from the upper end (of discharge), and the cumulated values are plotted against the appropriate discharges on a log normal graph. Figure 5.4 is an example. Because the percentages of days are cumulated from the high-discharge end, any point on the duration curve shows the percent of days (during the period used) that had discharges greater than the indicated value.

Although the duration curve is a cumulative frequency curve it should be interpreted merely as an expression of what happened in a particular time period, not as a probability curve because daily discharges are not only serially correlated but their characteristics change throughout the year. Consequently the probability of a daily discharge being greater than some specified value differs from that given by the duration curve because it depends both on the preceding flows and on the time of year.

The duration curve can be interpreted as follows: a daily flow may be expected to equal or exceed Q cfs P percent of the total time in a period of years equal to the period used in defining the duration curve. This interpretation would be reasonable only if the duration curve is based on tens of years of record. Duration curves based on one or only a few years of record may deviate greatly from the curve based on a long record.

Duration curves may be used to display graphically the variability of flow, a change due to stream regulation, or the dependability of low flows. An annual duration curve may be used with a sediment-rating curve to compute total annual sediment load. Discharges at 95 percent on the long-term duration curve are

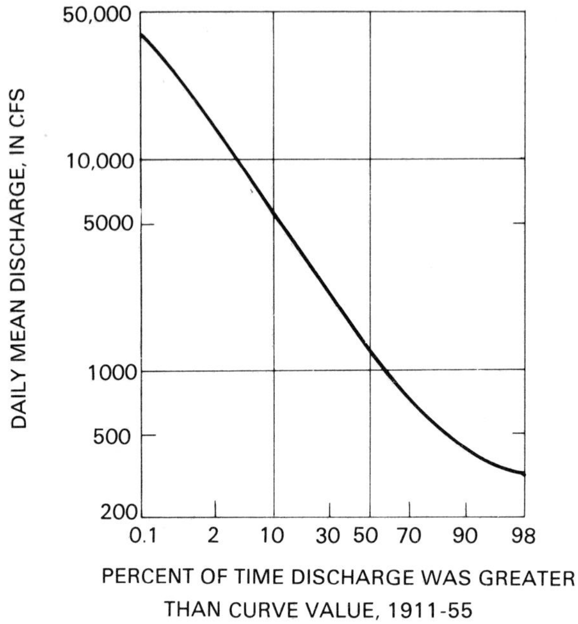

Fig. 5.4. Flow-duration curve for James River at Buchanan, Virginia.

related to points on the low-flow frequency curve but these relations change geographically.

5.6 BASE-FLOW RECESSION CURVES

Streamflow varies not only because the causative storms are random in their occurrence, geographical extent, duration, and intensity but also because various portions of the precipitation appear as measured streamflow at different times. The lag between precipitation and runoff depends on the path which the water follows from the surface of the ground to the stream. That portion which flows on the surface of the ground to the stream channel is most closely related in time to precipitation. Another portion may travel through the upper layers of the soil and arrive at the stream channels rather promptly but somewhat later than surface runoff. The surface flow and the shallow subsurface flow are often considered together under the title of direct runoff, which is the major portion of streamflow during floods. A third portion of the precipitation seeps downward into the ground-water aquifer that feeds the stream. This third portion is the last to reach the stream and is the portion that maintains streamflow during periods of no rain. It is commonly called the base flow of the stream.

The hydrograph of streamflow during periods when all discharge is derived from ground water sources is known as a base-flow recession. A curve which averages these recessions is the base-flow recession curve.

5.6.1 Theory

Theoretical equations for flow in an aquifer are the basis for equations describing a streamflow recession curve (Hall, 1968). These equations are generally of the form

$$Q_t = Q_o K^{-t} \tag{1}$$

where Q_t is the discharge at any instant, Q_o is the discharge at some initial time, K is the recession constant, and t is the time interval between Q_o and Q_t. The plot of log Q_t against t is a straight line on semilog paper. Rorabaugh (1964) showed that equation 1 is not correct shortly after a recharge. His equation for the movement of ground water, adapted from a heat-flow equation, has many terms, all but the first of which become very small after sufficient time has elapsed. After that critical time, which may be weeks or months depending on the aquifer characteristics, Rorabaugh's equation gives results comparable to equation 1. The Rorabaugh equation applies to one aquifer that is uniform, isotropic, and homogeneous; has a horizontal water surface; and has other specific boundary restrictions. Natural aquifers do not conform to these restrictions. In addition, a stream may be fed by several aquifers, each having different discharge characteristics and different rates and times of recharge. Furthermore, the aquifer or aquifers may lose water through evapotranspiration, and the water discharged to the stream is subject to variable evapotranspiration withdrawals daily and seasonally so that the recession of streamflow is not necessarily at the same rate as the recession of aquifer outflow. Other factors such as changes in channel storage further modify the streamflow recession. Consequently, the theoretical straight-line semilog recession curve is not applicable to many streamflow recessions.

Suppose a basin has a large-capacity artesian aquifer which releases water slowly to the stream and which is recharged only in unusually wet years. Suppose this basin also has a water-table aquifer which is recharged seasonally. Assume initially that both aquifers are recharged to capacity at the same time and that the individual recessions are represented by the solid lines beginning at zero days in Figure 5.5. Then the total base flow of the stream will be the sum of the two recession curves. This is represented by the dashed line which becomes tangent to the solid line of flatter slope as the contribution from the other aquifer approaches zero. A subsequent recharge only of the water table aquifer at 60 days will produce a combined recession curve of different slope and shape. Thus, base flow recessions of streams fed by more than one aquifer may take a variety of shapes and slopes at different times.

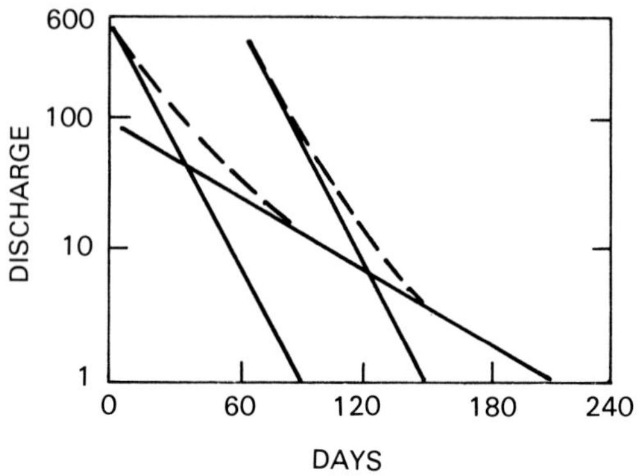

Fig. 5.5. Assumed recession curves (solid lines) from two aquifers and the streamflow recessions (dashed lines) which are the sums.

5.6.2 Derivation

Base-flow recession curves are commonly derived from segments of discharge hydrographs when there is no surface runoff in the channels. Weather records should be utilized in judging which segments are unaffected by precipitation. After the end of a storm sufficient time is allowed for the direct runoff and drainage from channel storage to have passed the gage. This time may range from a few days for a small stream to several weeks for a large one. Ordinarily the segment selected should be at least 10 days long because shorter segments may not be representative. And a segment should be steep enough so that subsequent discharges do not fall below an extension of that segment. Usually several years of hydrographs are needed to identify an adequate number of recession segments.

These selected segments can be synthesized into a full recession curve on a piece of tracing paper having discharge and time scales identical to those of the hydrographs. After the first segment is traced, the sheet is shifted horizontally so that the other segments fall into a common and continuous alignment. Figure 5.6 shows the segments traced on the sheet. A mean recession curve can be drawn through these.

Streamflow may recede at different rates in different seasons because of changes in evapotranspiration. The graphical synthesis described above will produce seasonally different recession curves only if the year is divided a priori. A method without this limitation begins with selection, from the identified recessions, of segments of a common length, such as 10 days. For each

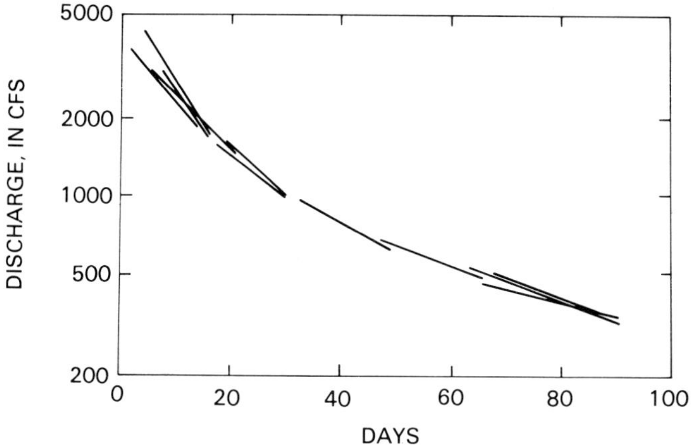

Fig. 5.6. Segments of hydrograph recessions traced on a sheet to define the recession curve, James River, Virginia.

segment, the discharges of the beginning day and the tenth day following are plotted as shown on the left graph of Figure 5.7. The points can be identified

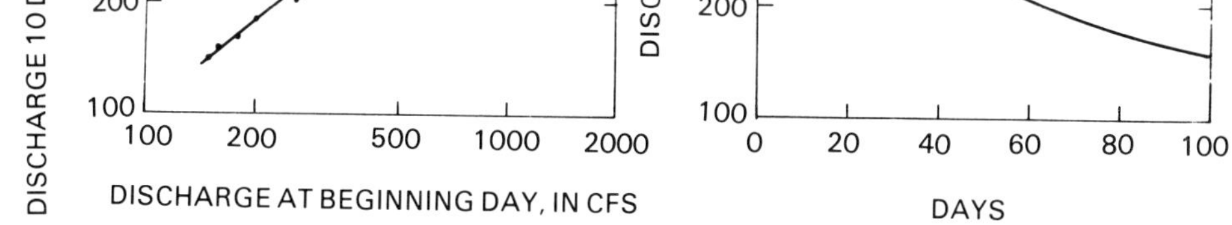

Fig. 5.7. Synthesis of a base-flow recession curve from selected 10-day hydrograph segments, Buffalo River near Lobelville, Tennessee.

by month so that mean relations by seasons can be drawn if significant differences are found (they were not found in these data). The mean line of figure 5.7 was transformed to the recession curve by beginning at 1000 cfs plotted at zero days. Then the discharge 10 days later is plotted at day 10. Next, the day 10 discharge is used as beginning discharge to determine the discharge 10 days later (to be plotted at 20). This process is repeated until the complete recession curve (the right curve of Figure 5.7) is defined.

5.6.3 Assumptions as to recharge

Most substantial rises in streamflow are accompanied by an apparent increase in base flow which may be drainage from bank storage rather than from a recharge to the aquifer. The assumptions made as to recharge, or lack of it, from a particular storm in a particular season greatly influence the characteristics of the derived recession curve and consequently its reliability and meaning. The appropriate assumption as to recharge depends on the basin and channel characteristics, the magnitude of the precipitation, antecedent conditions, and the time of year. Knowledge of the basin and study of the hydrograph in conjunction with weather records should lead to a reasonable concept for a particular stream site.

5.6.4 Seasonal variability

Evapotranspiration losses from the aquifer and from the channel vary greatly with time of year and may vary with elevation of the water table and with stream discharge. Consequently, it has been the practice to define separate recession curves for summer and for winter periods. The winter recession curve should more closely represent the discharge from ground-water sources because the losses to the atmosphere are minor during winter. Unfortunately, the frequency of precipitation during winter and the fluctuations of temperature above and below freezing produce such fluctuations in streamflow that it becomes nearly impossible to discriminate periods of base flow for many streams. Consequently, most recession curves are defined for summer conditions. Data available for Brandywine Creek at Chadds Ford, Pa. permit the definition of curves for both summer and winter (see Figure 5.8). Obviously, the recession rate does not change abruptly from one curve to the other; therefore these curves do not apply to spring and fall conditions. Other factors that cause variable recession rates are described by Riggs (1964b).

5.7 WATER QUALITY CHARACTERISTICS

The great many constituents and characteristics of a natural water rule out any simple description of the quality of a water. But some generalizations are useful as indicators of whether a water is likely to be suitable for a particular purpose. The significant characteristics of water and its impurities that affect quality for various uses are discussed by Camp (1963) and by Hem (1972, p. 320-336). These characteristics include temperature and the amounts of suspended sediment, dissolved solids, and biological material in the water.

Water temperature changes seasonally; it may even change throughout a day on small streams. The variation of weekly or monthly mean temperatures throughout the year provides information needed for many uses. Figure 5.9 shows the seasonal cycles of water temperature in a northern and a southern State.

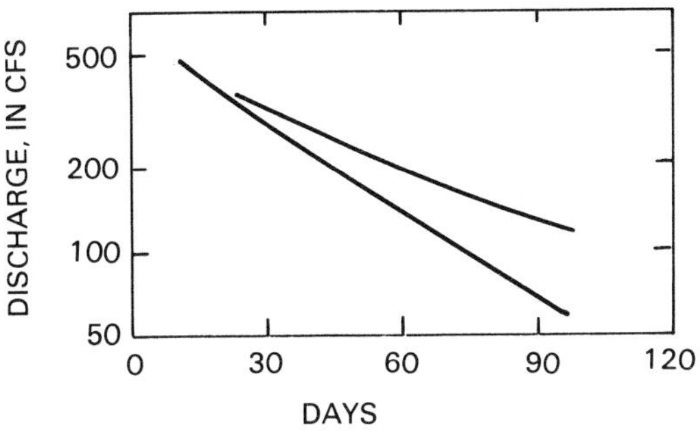

Fig. 5.8. Winter (upper) and summer (lower) base-flow recession curves for Brandywine Creek near Chadds Ford, Pennsylvania.

The sediment transported by a stream is a measure of water quality, and it is the basis for computation of the sediment deposition to be expected in reservoirs, downstream channels, or estuaries. Sediment concentration is needed in the design and management of water-treatment facilities; it can be characterized by the frequency distribution of daily means. Where only occasional sediment measurements are available the distribution of daily mean concentrations can be approximated from a sediment-rating curve (Section 3.1.9) and a flow-duration curve.

The mean annual sediment load is the appropriate characteristic for use in estimating the amount of sediment that will be deposited in a reservoir. A short record of annual sediment load may be extended by use of a long streamflow record and a sediment rating curve.

Chemical quality of a water depends on the amounts of each of many constituents (Swenson and Baldwin, 1965). The water may contain minerals and organic compounds that are undesirable for certain uses. Mineral content also determines whether the water is hard or soft and acid or alkaline.

Specific conductance is an indication of total dissolved solids. Waters having specific conductances of less than a few hundred micromhos per centimeter generally are considered good. But the concentration of dissolved solids, and thus the conductance, changes with discharge. A more informative characteristic than the maximum or the average conductance would be the frequency distribution of daily mean conductances or the relation of conductance to stream discharge.

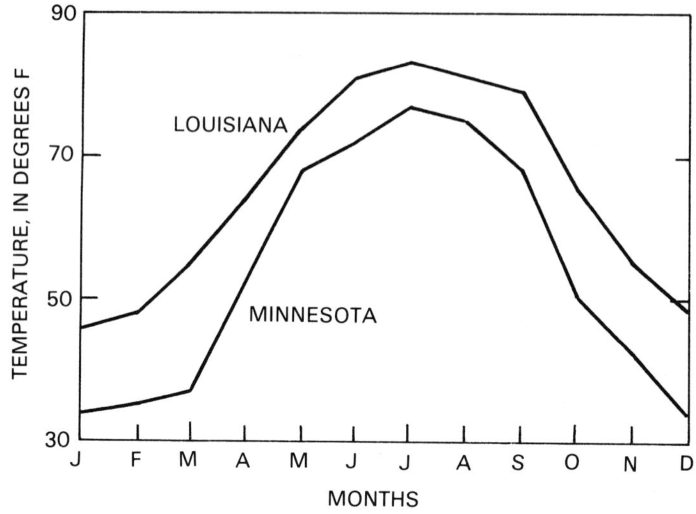

Fig. 5.9. Variation in mean monthly water temperatures.

Various methods of graphically representing the concentrations of the principal dissolved constituents are described by Hem (1972). Such graphs permit ready comparisons among waters.

Hardness is indicated by the concentration of calcium carbonate or its equivalent; water with less than 60 mg/l of hardness is considered soft. The pH is the characteristic that indicates whether the water is acid or alkaline and to what degree.

The biological content of a water is usually the result of pollution and cannot be simply characterized quantitatively, although the dissolved oxygen content is an indicator.

Simplistic indicators of chemical quality are useful for preliminary studies and evaluations but more detailed information is usually required in planning a specific use of a water.

REFERENCES

Camp, T.R., 1963, Water and its impurities: New York, Reinhold Publ. Corp., 355 p.

Carroon, L.E., 1970, Correlative estimates of streamflow in the upper Colorado River basin: U.S. Geol. Survey Water-Supply Paper 1875, 145 p.

Crawford, N.H. and Linsley, R.K., 1966, Digital simulation in hydrology: Stanford Watershed Model IV: Dept. of Civil Engineering, Stanford University, Tech. Rept. No. 39, 210 p.

Fiering, M.B., 1963, Use of correlation to improve estimates of the mean and variance: U.S. Geol. Survey Prof. Paper 434-c.

Hall, F.R., 1968, Base-flow recessions - a review: Water Resources Research, Vol. 4, No. 5, p. 973-983.

Hem, J.D., 1972, Study and interpretation of the chemical characteristics of natural water: U.S. Geol. Survey Water-Supply Paper, 1473, Second Edition, 363 p.

Martin, R.O.R., 1964, Use of precipitation records in the correlation of streamflow records: International Assoc. of Scientific Hydrology Bull., Vol. IX, No. 4, p. 24-31.

Moore, D.O., 1968, Synthesizing daily discharge from rainfall records: Journal of Hydraulics Division, ASCE, Vol. 94, No. HY5, Proc. Paper 6119, p. 1283-1298.

Riggs, H.C., 1964a, Stream discharge regressions using precipitation: U.S. Geol. Survey Prof. Paper 501-C, p. 185-187.

Riggs, H.C., 1964b, The base-flow recession curve as an indicator of ground water: International Assoc. of Scientific Hydrology Publ. 63, Berkeley, p. 352-363.

Rorabaugh, M.I., 1964, Estimating changes in bank storage and ground-water contribution to streamflow: International Assoc. of Scientific Hydrology Publ. 63, Berkeley, p. 432-441.

Schneider, W.J., 1961, Precipitation as a variable in the correlation of runoff data: U.S. Geol. Survey Prof. Paper 424-B, Article 9.

Sittner, W.T., Schauss, C.E., and Monro, J.C., 1969, Continuous hydrograph synthesis with an API-type hydrologic model: Water Resources Research, Vol. 5, No. 8, p. 1007-1022.

Swenson, H.A. and Baldwin, H.L., 1965, A primer on water quality: U.S. Geol. Survey, 27 p.

Chapter 6

RELATION OF GROUND WATER TO STREAMFLOW

6.1 INTRODUCTION

Ground water and surface water are often studied independently but a particular particle of water may be considered ground water at one time and surface water at another in its transit of the land. A proper interpretation of streamflow characteristics requires an understanding of how ground water is replenished and depleted and how these changes relate to streamflow.

Ground water fills the pore spaces and fractures in the earth materials to a level known as the water table. The porous material that contains water is called an aquifer. If the porous material extends to the ground surface, so that the water table is free to move vertically in response to changes in volume of water in storage, the aquifer is called a water-table aquifer. If the upper part is confined by an impermeable or semipermeable rock formation, and the ground water is under pressure, the aquifer is artesian.

Many aquifers in humid regions discharge at least for part of the year to the ground surface either through springs or directly to a stream channel as shown in Figure 6.1. Kilpatrick (1964) showed that the low flow of a small stream in Georgia was derived from discharges of a water-table aquifer and from upward leakage from an artesian aquifer. Aquifers in arid regions may discharge only by evapotranspiration.

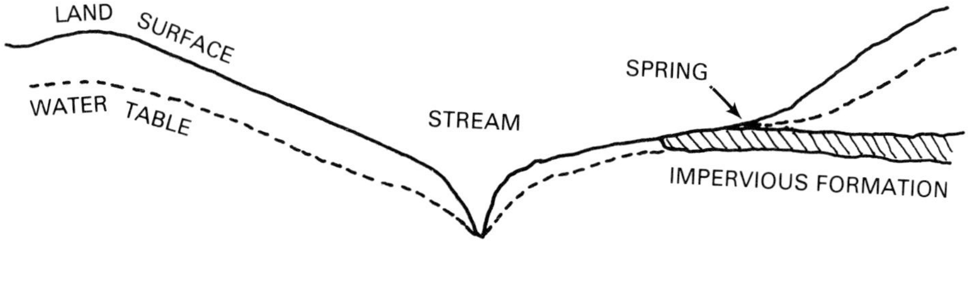

Fig. 6.1. Valley cross section showing how aquifers discharge to a stream and to a spring.

The rate of movement of ground water depends on the hydraulic gradient and the transmissivity of the aquifer material. Transmissivity may range over more than 2 orders of magnitude. Consequently the velocity of ground water may range from a few feet per year in clay to many feet per day through very porous material. Soluble rocks, such as limestone, often contain solution channels through which ground water moves rapidly. Figure 6.2 shows hydrographs of 2 springs which illustrate the different rates of ground-water movement. Jack Daniel Spring drains a limestone area; its rapid response to rainfall indicates the probable existence of sinks and solution channels in the basin. The more typical hydrograph of Indian Ford Spring reflects slow infiltration through the soil and slow movement through the aquifer; it does not follow the hydrograph of the nearby stream.

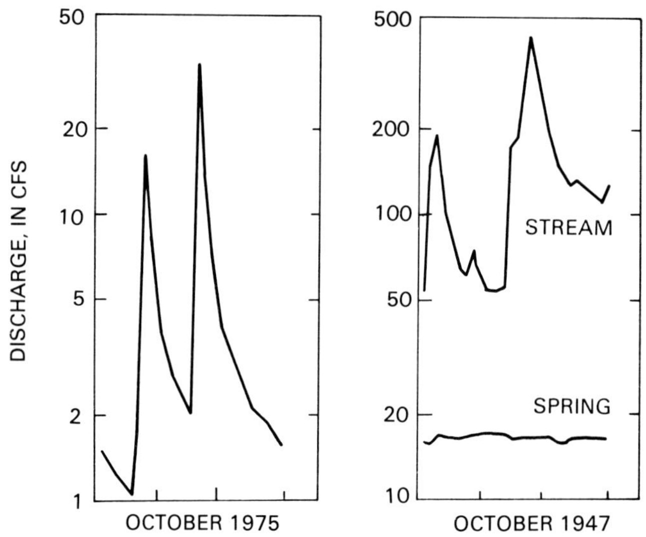

Fig. 6.2. Hydrographs of two springs, and of a stream near one of them, showing the differences in flow variability.

Spring flow is less variable if the recharge area is distant from the discharge point and if the aquifer is not subject to evapotranspiration. Such springs are better indicators of ground-water discharge than are the low flows of streams.

6.2 AQUIFER RECHARGE

Ground water is replenished by rain or melting snow which infiltrates the soil. In temperate, humid climates the major replenishment or recharge occurs during the winter and spring months when precipitation is plentiful, evapotranspiration is low, and soil moisture is high. During the growing season a large part of the precipitation is held in the soil until it is returned to the atmosphere by evapotranspiration. Thus the contribution of a substantial rain to the ground water body ranges from a maximum in winter or early spring to very little following a long dry period in summer.

In most arid regions the water table is below the streambeds and thus the streams are ephemeral. Here a substantial storm will produce streamflow and some infiltration over the basin but little of that infiltrated water gets below plant roots. Recharge, if any, occurs as leakage through the streambeds whose permeability may be increased temporarily by the fast, turbulent flows. Jordan (1977) reported substantial transmission losses in streams in western Kansas; most of the water lost from these channels is accretion to ground water.

The amount of recharge to an aquifer from a storm may be computed from the rise in the water table as measured in wells, the estimated porosity of the aquifer material, and the extent of the aquifer. Porosity is the percentage of void space relative to the total volume of the mass.

6.3 HYDROGRAPH INTERPRETATION

The streamflow hydrograph contains considerable information about the hydrology of a basin. Highly-variable daily discharges indicate rapid runoff and little infiltration, especially if the mimimum flows are particularly low. Very uniform flows throughout the year are produced by basins on which much of the precipitation reaches the water table on its way to the stream.

The hydrographs of Figure 6.3 contrast the runoff patterns of drainage basins only 50 miles apart. Runoffs in inches are about the same but the annual

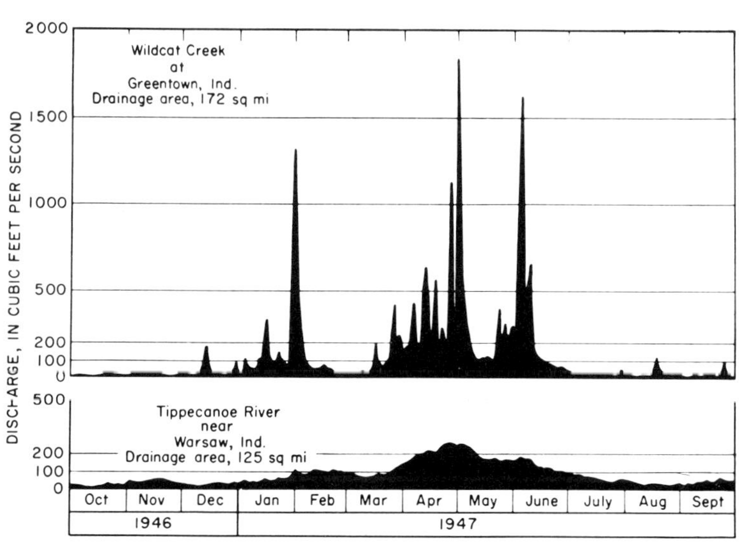

Fig. 6.3. Contrast in runoff patterns of streams draining basins of different permeability (From McGuinness, 1963).

variability as well as the daily variability is much greater on Wildcat Creek. According to McGuinness (1963) Wildcat Creek flows from a basin floored with clayey till. The creek responds quickly to precipitation and then falls off rapidly to a low base flow. Tippecanoe River drains a basin in much of which permeable sandy gravelly glacial outwash lies at the surface. This outwash absorbs much of the precipitation, preventing sharp flood peaks and releasing water slowly to maintain a larger base flow.

The relation of ground water to streamflow can be implied by study of the stream hydrograph. At the end of a long period of fair weather the flow in a stream usually will be from ground water (but it will be less than the discharge from the ground-water body because of evapotranspiration). Overland flow from a heavy rain will increase the flow in the stream channel as shown in Figure 6.4.

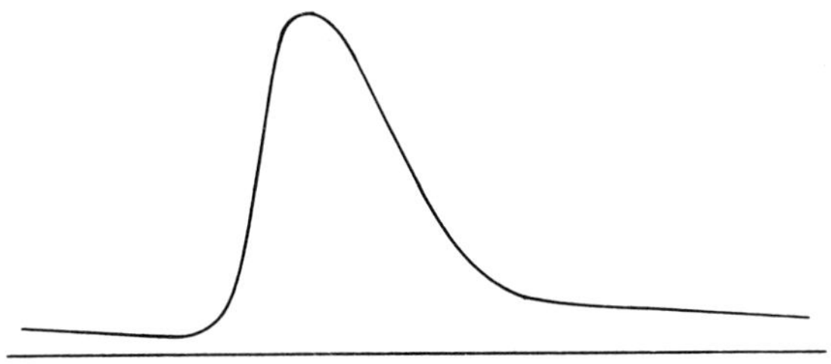

Fig. 6.4. Streamflow hydrograph showing the increase in base flow (Ground water discharge) resulting from a storm.

Some of the water that arrives rather quickly in the channel travels by a shallow subsurface course; it may be regarded as ground water although it belongs in a different category than water in the formations below the water table. Another portion of the rainfall infiltrates to the ground water body, raising the water table, increasing the head toward the stream, and thus increasing the base flow rate over that prior to the storm as shown in Figure 6.4.

The portion of streamflow derived from ground water during or closely following a period of overland runoff cannot be estimated closely. However water-table profiles near a stream during changes in stream stage indicate the direction of ground-water flow at various times. As shown in Figure 6.5, the gradient of the water table is towards the stream when the stream stage is low. As the stage rises, the water-table gradient decreases and finally reverses, at which time water from the stream flows into the banks. At the same time ground water that otherwise would have gone to the stream is being stored further inland. As the stream stage recedes the water table gradient begins its return to normal.

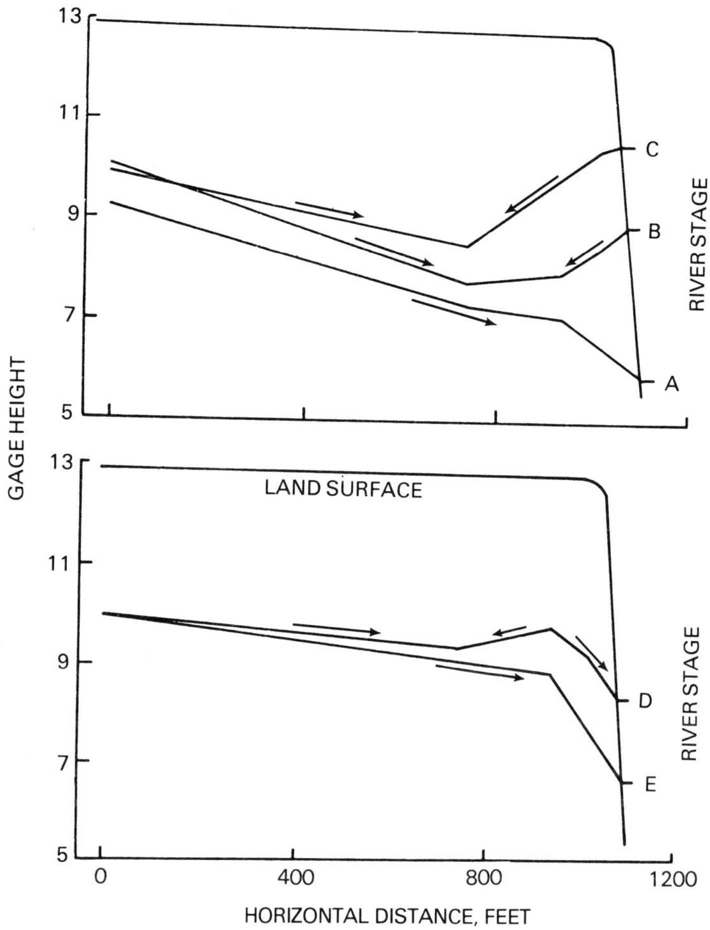

Fig. 6.5. Ground water levels and flow directions during the rising and falling stages of a stream (After Daniel and others, 1970).

Two conclusions can be drawn from this. First, the ground-water contribution to streamflow during a flood runoff becomes negative during the rising stage and later increases above the previous base flow as the stage decreases (Fig. 6.6). Second, the water stored near the channel during the rise takes some time to drain away, thus, for some time after the flood, the streamflow recession includes drainage both from bank storage and channel storage in addition to aquifer drainage.

In the example of Figure 6.5, the stream stage was not high enough to inundate the adjacent land surface. When major floods inundate large areas along their channels, the vertical infiltration is large and the resultant storage in the flood plain may contribute to streamflow for a long period.

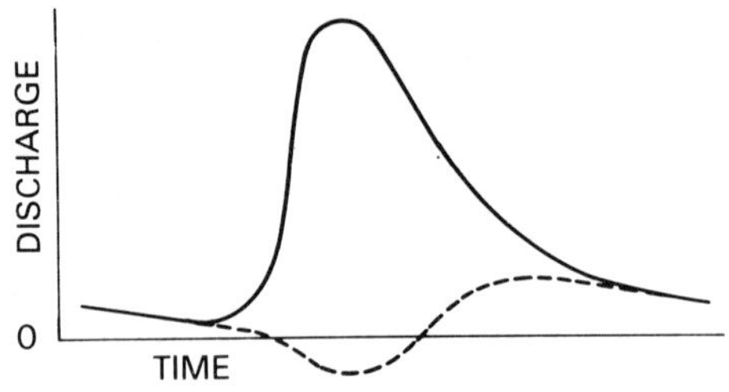

Fig. 6.6. Hydrograph showing ground water contribution to a stream during a rise in streamflow.

The above interpretation of the streamflow hydrograph during and following a flood is helpful in identifying those segments of a hydrograph which represent only base flow. Such segments are used to define base-flow recession curves (Section 5.6) and to estimate the ground-water yield to a stream (Riggs, 1963). Obviously the recession shortly after a hydrograph rise may include drainage from bank storage but whether a recharge of the principal aquifer(s) was associated with a rise may be less clear. A hydrograph like that of Figure 6.4 does not always indicate that a recharge occurred. In Figure 6.7, the hydrograph for the latter part of July and the first half of August defines the slope of the base-flow recession curve. An extension of that curve through September would be very close to the hydrograph for most of that month, indicating that the storm of late August did not recharge the aquifer appreciably.

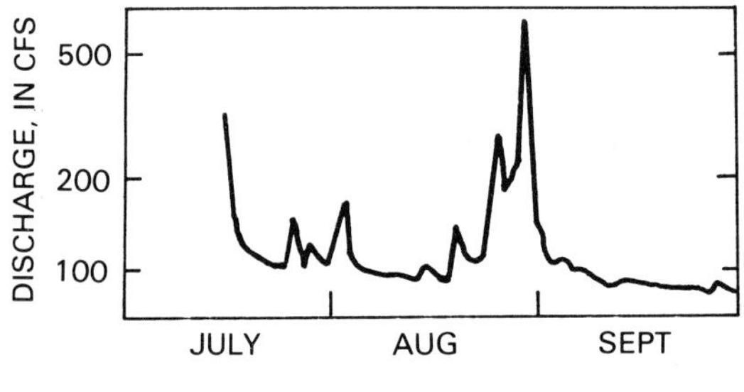

Fig. 6.7. Hydrograph for N.F. Obion River, Tennessee, which indicates negligible recharge from the late August storm.

The assumption that recharge occurred from a storm that produced appreciable direct runoff can be checked. Consider Figure 6.8 in which the recessions (dashed) indicate appreciable recharge in May and June. The volume of recharge is the difference in the volumes under the two recession curves from some common date to such time as the curves reach negligible discharges. The Red River

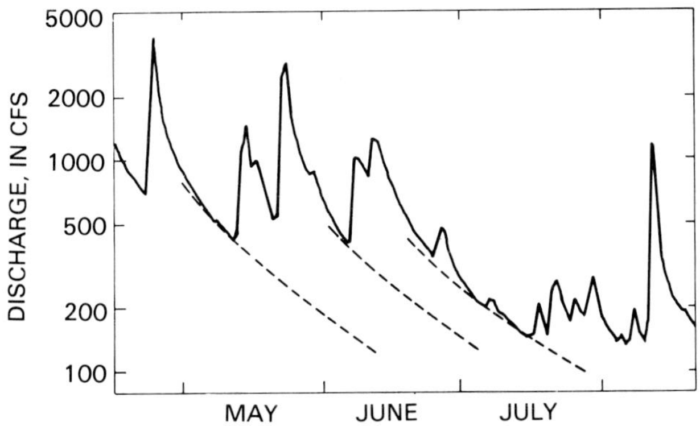

Fig. 6.8. Hydrograph for Red River, Tennessee, 1955, showing recession curves which indicate recharge from May and June storms.

recession curve would drop from 500 cfs to 1 cfs in about 150 days. The volume difference can be defined graphically - or mathematically if the equation of the recession curve is defined. The computed recharge should be reasonable with respect to the rainfall, the measured runoff, and the estimated evaporation. If it is not, the estimate of the amount of base flow following (or preceding) the rise is incorrect. In the absence of rainfall records one might appraise whether the computed recharge is reasonable for the magnitude of the rise in streamflow. Data for other rises would be needed for this test; see references in section 11.5.3.

A word of warning; the storm hydrograph of an ephemeral stream may look similar to one for a stream with a dependable ground-water contribution. Study of such a hydrograph for a few months will show that the apparent increase in base flow following a peak is not due to recharge of a ground-water body; it probably reflects the drainage from channel and bank storage.

6.4 BANK STORAGE IN SURFACE RESERVOIRS

The storage capacities of surface reservoirs at various levels are defined by topographic surveys. The resulting stage-capacity curves are used in reservoir operations. But when some reservoirs are filled for the first time they take considerably more water than is indicated by the stage-capacity curve. And later, a reduction in reservoir stage produces more water than expected. These effects are due to the infiltration of water into the banks of the reservoir (presumably the project investigation assured that the geologic setting contained nothing that would permit stored water to bypass the dam).

Bank storage below the operating stage of a reservoir is merely an addition to dead storage. But within the operating range bank storage is increased during rising stages and is decreased during falling stages. The magnitude of bank storage changes may be large enough to be taken into account in reservoir operation. For example, Simons and Rorabaugh (1971) computed the dead or inactive aquifer-storage capacity of Hungry Horse Reservoir, Montana as about 50,000 cfs-days, and the active capacity about 100,000 cfs-days. However, the available water from aquifer storage depends on the magnitude of the stage reduction and on the time the stage is held at the lower level; complete drainage from a bank aquifer may take months.

6.5 THE WATER RESOURCE

Although there are many reasons for studying ground water and surface water separately, one should recognize that the total renewable water resource of a basin in the flow of the stream as it leaves the basin. The ground-water resource is included in the streamflow; any withdrawal of ground water will reduce streamflow. But the two resources are independent, or nearly so, in regions where the water table is below stream channels and is not affected by infiltration from the streams. Here only the surface water resource is renewable; the ground water resource is not.

Within a basin it is possible to provide a more dependable supply by utilizing storage of water in the ground if the storage aquifer is not closely connected (in time) to a stream. Then withdrawals of ground water will not reduce streamflow during the low-water period. Conjunctive use is a management tool to supplement low streamflows; it does not change the total supply.

Other activities of man in which ground water needs to be considered are channel deepening, which may provide a better connection to an aquifer; raising a river level by dams, which raises the level of the adjacent water table; diversion of streamflow from a losing stream, which may reduce the downstream ground-water recharge; and routing of reservoir releases down a channel where the water table is in continuity with the stream. Moench, Sauer and Jennings (1974) found that results of routing reservoir releases were appreciably improved by inclusion of a gross simplification of the ground-water system.

REFERENCES

Daniel, J.F., Cable, L.W., and Wolf, R.J., 1970, Ground water-surface water relation during periods of overland flow: U.S. Geol. Survey Prof. Paper 700-B, p. 219-223.

Jordan, P.R., 1977, Streamflow transmission losses in western Kansas: Jour. of the Hydraulics Division, ASCE, Vol. 103, No. HY8, p. 905-919.

Kilpatrick, F.A., 1964, Source of base flow of streams: International Assoc. of Scientific Hydrology Publ. No. 63, Berkeley, p. 329-339.

McGuinness, C.L., 1963, The role of ground water in the national water situation: U.S. Geol. Survey Water-Supply 1800, 1121 p.

Moench, A.F., Sauer, V.B., and Jennings, M.E., 1974, Modification of routed streamflow by channel loss and base flow: Water Resources Research, Vol. 10, No. 5, p. 963-968.

Riggs, H.C., 1963, The base-flow recession curve as an indicator of ground water: International Assoc. of Scientific Hydrology Publ. No. 63, Berkeley, p. 352-363.

Simons, W.D. and Rorabaugh, M.I., 1971, Hydrology of Hungry Horse Reservoir, northwestern Montana: U.S. Geol. Survey Prof. Paper 682.

Chapter 7

FLOW CHARACTERISTICS AT UNGAGED SITES

7.1 INTRODUCTION

Although hydrologic data have been collected at thousands of sites, those sites constitute only a small sample of the sites where flow characteristics may be needed. The information at gaged sites permits the calibration of various procedures for transferring flow characteristics to ungaged sites or estimating them from rainfall. Methods are categorized into ones that require drainage-basin and climatic characteristics obtainable from maps and climatic records, and those that require a small amount of data to be collected at the site of interest.

7.2 NO DATA AT SITE

7.2.1 Regression analysis

Regional analysis of streamflow characteristics is a widely applicable and fairly simple method. Using data at gaging stations, a flow characteristic is related to basin and climatic characteristics, usually by multiple regression. An example is

$$\log Q_{25} = -2.07 + 0.97 \log A + 2.11 \log P \tag{1}$$

where Q_{25} is the 25-year flood in cfs in the Snohomish River basin, Washington; A is drainage area in square miles; and P is mean annual precipitation in inches. The equation is in a linear form, a requirement of the usual regression method. It may be transformed to

$$Q_{25} = 0.0085\, A^{0.97}\, P^{2.11} \tag{2}$$

For a stream site in the Snohomish River basin Q_{25} can be estimated from the equation using drainage area measured above the site, and the mean annual rainfall on that drainage area from an isohyetal map. The reliability of the result is indicated by the standard error of the regression, 0.14 log units, which corresponds to +38 and −28 percent.

Multiple regression is directly useful as a regionalization tool because the discharge for a given frequency level can be related to basin characteristics, leaving residuals that may be considered as due to chance. The regression line averages these residuals. Thus, in one operation, the effects of differing basin characteristics are preserved and the chance variation is averaged.

The regression model used should approximate the physical process as closely as possible and it must be in a linear form. The log-linear form is suitable for many hydrologic relations but it should not be accepted blindly. Graphical multiple regression is a good way to define the suitable model; a graphical solution need not be linear, however. Figure 7.1 shows the graphical solution using the data on which eq. 1 was based. This confirms that the relations with A and P are log linear and that the form of eq. 1 is appropriate for these data. The equation of the graphical relation of Figure 7.1 is virtually the same as equation 1.

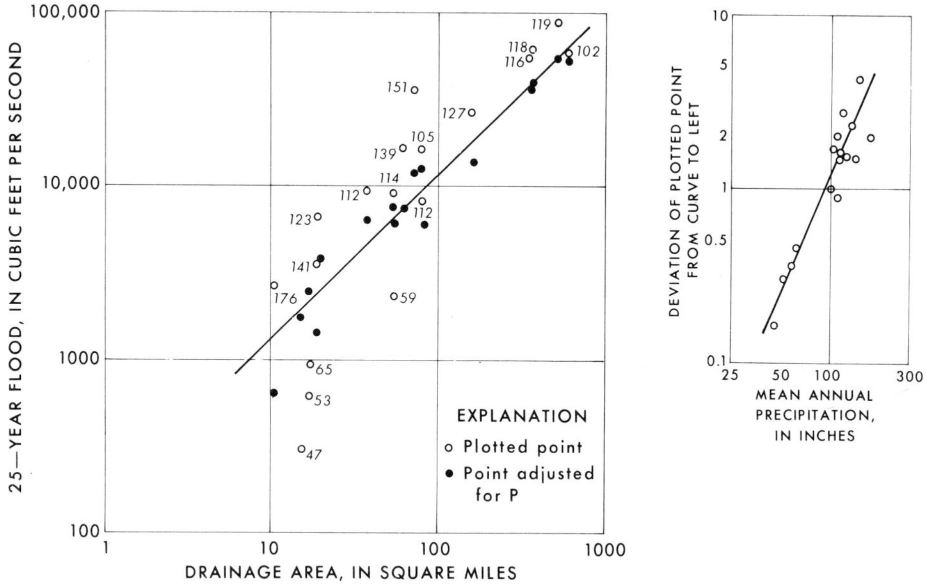

Fig. 7.1 Graphical analysis of data used to define equation 1.

If a graphical multiple regression is developed on arithmetic graph paper, rather than on log paper, and is found to be linear the model is

$$Y = a + b_1X_1 + b_2X_2 \tag{3}$$

and the standard error is in units of Y. Standard error using this model is sometimes stated in percent of mean Y; the percentage increases with distance from the mean. See section 4.4.1.

In model equation 3 the effects of the variables are additive whereas in the model used in equation 1 the effects are multiplicative. If a variable can take a value of zero then the log-linear model cannot be used with that variable because zero does not have a real logarithm. This limitation can be overcome by adding a small constant, commonly 1 or 0.1, to all values of the variable before using the variable in regression. Before doing this the analyst should assure himself that the addition will not violate the concepts used.

The independent variables selected for use in the regression analysis should be ones that can be readily quantified at ungaged sites and that are postulated to have a particular effect on the dependent variable (the flow characteristic). Selection of independent variables is often made on a statistical basis; that is, many variables are used in preliminary regressions and those that lack statistical significance are discarded. This approach occasionally results in retention of a variable whose effect on the dependent variable does not conform to hydrologic principles. Usually the effect of such a variable is trivial. It seems better to select in advance those variables which are expected to have practical significance - to rely on hydrologic judgment rather than on statistical selection. However, a variable that is known to affect a flow characteristic in a particular way may not be statistically significant in a regression if the range of that variable in the sample is small. Channel slope as an estimator of flood characteristics is an example of a variable that has little range in some regions.

Some variables such as basin storage, channel slope, and storm magnitudes are appropriate for estimating flood characteristics but have little effect on mean flow. Mean flow in humid regions of homogeneous geology is closely related to drainage area and mean annual precipitation.

Not all important variables can be quantified from maps. It is sometimes desirable or necessary to use surrogates. Occasionally the regression coefficient of a variable selected on a hydrologic basis appears to represent something other than that intended. For example, consider the relation of 50-year flood to drainage area and percent of area forested,

$$P_{50} = 6.03\ A^{0.81}\ F^{0.84}$$

for California mountain streams. For this region of large topographic relief the regression indicates that an increase in percent area forested will produce an increase in 50-year flood, contrary to expectation. But examination of maps shows that forest cover increases with elevation, and presumably with precipitation. Thus forest cover may be a surrogate for precipitation. Although surrogate variables are useful, extrapolation of a regression equation which includes surrogates may produce questionable results.

The applicability of a regression equation to ungaged streams in the region for which it was derived depends on how well the data used in its derivation cover the ranges of variables in the region, and whether the variables used are adequate descriptors of the flow characteristic at all of the streams to which the regression equation might be applied.

If more than one regression equation is derived for a single flow characteristic, the equation with the smallest standard error is commonly used. However, one with fewer variables or with only those variables that are easy to quantify at an ungaged site may be selected if its standard error is not appreciably greater than the smallest standard error obtained. There may be no practical difference in the results from two equations whose standard errors differ considerably.

Only equations in which all the regression constants are statistically significant should be considered. Computer programs for regression analysis either produce regression equations which include only those variables whose effects are statistically significant at the specified level (usually 95%), or they show the statistical significance of each variable in the equation.

Regression on basin characteristics produces best results in humid regions of fairly uniform terrain and gradual areal changes in precipitation. The method is not generally useful for estimating low flows because of the large influences of basin geology and evapotranspiration on those flows. Among the factors that limit the method are (1) large interchanges of surface and ground water due to geologic conditions, (2) large ranges in elevation in the basins, and the corresponding large range in basin precipitation, (3) poor definition of precipitation in arid regions, and (4) flow modification due to man's activities.

Regression equations and their standard errors are given for many flow characteristics in four regions of the United States by Thomas and Benson (1970).

7.2.2 From rainfall

Techniques for estimating flood-peak characteristics from rainfall are discussed in Chapter 8.

Rainfall ordinarily is not a major indicator of low-flow characteristics.

Mean flows of ungaged streams may be approximated from an isohyetal map if (1) flow records from some basins in the region are available to develop a calibration, (2) the isohyetal map is based on precipitation records (some isohyetal maps are based partly on streamflow records) and (3) the geology of the region is more or less homogeneous. In regions of high relief and the consequent high range in precipitation from the upper to the lower part of the basin, a runoff-altitude relation can be developed by trial-and-error from gaged records, even in the absence of precipitation records (Riggs & Moore, 1965). This relation can be used to estimate mean flow from the measured areas in each of several elevation ranges of an ungaged basin; the sum of these partial runoffs is the runoff from the ungaged basin.

7.2.3 Interpolation along a channel

Interpolation of flow characteristics between gaged points on a stream usually produces more reliable results than regression on basin characteristics. The flow characteristics are plotted against channel length (Fig. 7.2); points of interest can be readily identified on the graph by their relation to a gaged point. A method of interpolating low flows is given in Chapter 9. See Chapter 8 for another example of the method applied to floods.

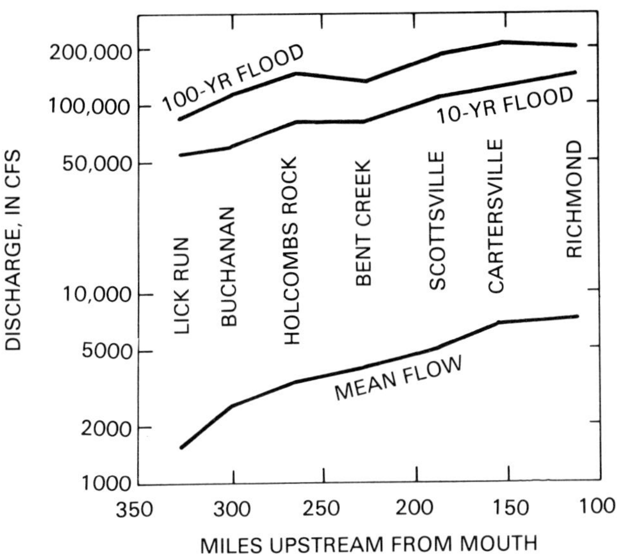

Fig. 7.2. Flow characteristics along the main stem of James River, Virginia.

7.3 SOME DATA AT SITE

Flow characteristics in some regions are not closely related to variables that can be obtained from maps, the usual variables cannot be quantified reliably, or a simple regression model does not explain the relation. In arid and semarid regions precipitation is highly variable in time and space, the areal extent of storms is often limited, and weather records are few. In basins of large topographic relief, precipitation in the headwaters may be 4 or 5 times that in the lower reaches; a basin average would be a poor indicator of flow characteristics. When topographic and ground-water divides are not coincident the runoff from a basin may be more, or less, than that indicated by basin precipitation. For example, the annual runoffs of two adjacent streams in the headwaters of Umpqua River, Oregon, are 33 and 54 inches; the difference is largely due to interbasin movement of ground water. Some drainage basins include areas that do not contribute surface runoff or that contribute only occasionally. The contributing drainage area for such a basin cannot be defined reliably. And the efficiency with which a stream system transmits water depends both on geology and on topography neither of which can be adequately expressed by simple variables in some regions.

A consequence of one or more of these conditions is that mean flow or flood peaks on some streams decrease with increasing drainage area. A regression model cannot be devised that will apply to a region in which some streams exhibit this anomalous characteristic.

Flow characteristics of streams in the settings described above usually can be estimated more reliably from some information at the site than from information taken from maps or climatological records. Discharge measurements and measurements of channel geometry are the most useful site data.

7.3.1 Mean flow from monthly measurements

Where mean flow is not closely related to drainage area, Riggs (1969) has shown that reliable estimates in some regions can be made from monthly discharge measurements for a year. The procedure is demonstrated in the region shown in Figure 7.3. The plot of Figure 7.4 establishes that drainage area is not a useful estimator of mean flow in this region.

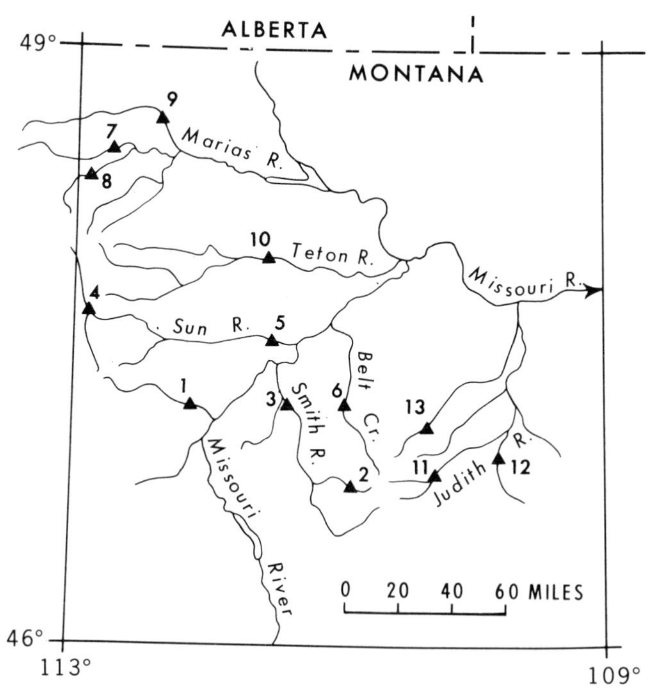

Fig. 7.3. Part of Missouri River basin showing locations of gaging stations.

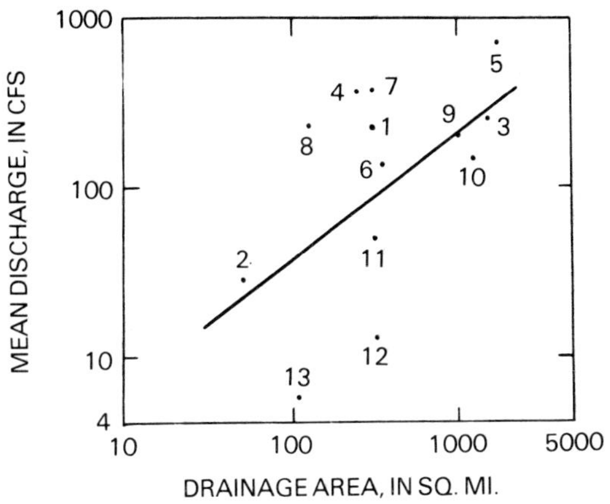

Fig. 7.4. Relation of mean flow to drainage area at gages shown in Figure 7.3.

Both an explanation and a verification of the method are provided by using data from two gaged streams, one of which, Two Medicine Creek (number 7) is considered ungaged. Cut Bank Creek (number 9) is used as the index station. Assume that the discharges of Two Medicine Creek on the 15th of each month are equivalent to what would have been measured on those days. The concurrent daily discharges and the monthly means of Cut Bank Creek would be available from the record. These three items of data are listed in Table 7.1. Although the objective of this technique is to estimate annual flow, this is obtained by estimating and summing monthly discharges.

The plot of concurrent midmonth discharges of the two streams in Figure 7.5 shows that the relation changes at least monthly. These changes are due to differences in the timing of snowmelt and to a pattern of monthly diversions from Two Medicine Creek, ranging from zero to over 10,000 acre-feet, which is reasonably consistent from year to year. Obviously a line averaging the points of Figure 7.5 would not be a reasonable basis for estimation.

The Riggs method uses a different relation of unit slope for each month, its position determined by the plotted point. The dashed line through point 10 on Figure 7.5 is the relation used for transferring the October monthly mean of Cut Bank Creek to get the Two Medicine Creek estimate. Graphical transfer is not necessary; the monthly mean flow of Two Medicine Creek usually is obtained by multiplying the measured monthly mean flow of Cut Bank Creek by the ratio of the midmonth discharges. Results are shown in Table 7.1 along with the measured values for comparison. Some of the monthly estimates are considerably in error but the annual mean is less than 6 percent from that measured.

TABLE 7.1 Data and results of computing annual mean flow from monthly measurements

Date	Daily mean discharge, cfs		Month	Monthly mean discharge, cfs		
	Two Medicine Creek	Cut Bank Creek		Cut Bank Creek	Two Medicine Creek	
					Computed	Measured
10/15/60	244	36	10	32.6	220	55.3
11/15	56	26	11	28.5	61	58.2
12/15	50	44	12	26.0	30	36.1
1/15/61	40	30	1	23.9	32	29.5
2/15	50	65	2	48.3	37	82.2
3/15	156	72	3	67.0	140	128
4/15	175	82	4	94.2	200	231
5/15	978	332	5	471	1040	1211
6/15	850	470	6	509	920	1010
7/15	82	164	7	148	74	114
8/15	50	74	8	58.6	40	22.6
9/15	85	42	9	40.7	82	60.5
Annual mean, 1961 water year				129	240	254

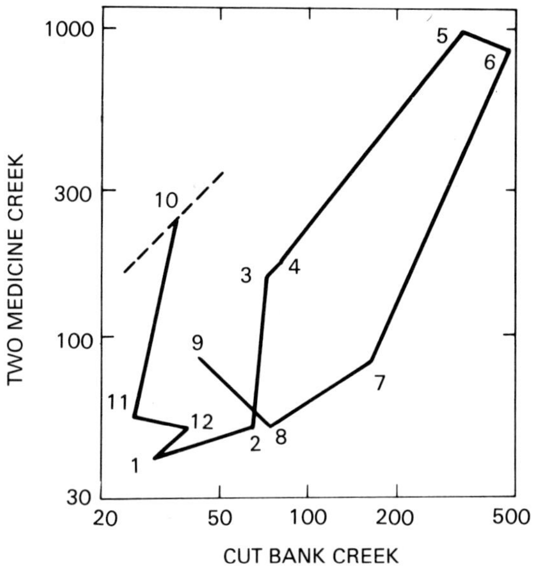

Fig. 7.5. Concurrent mid-month discharges, in cfs, from Table 7.1.

The final step in this procedure is to relate annual means for 1961 to means of record for streams in the area, to define a mean relation line (Figure 7.6) and to transfer the estimated 1961 mean of 240 through that relation to get an estimate of the long-term mean for Two Medicine Creek. The long-term mean of 430 cfs, as estimated by this method, compares favorably with 385 cfs based on 31 years of record.

This method is applicable to streams in regions where flow is principally from snowmelt or from ground water, or where storms cover large areas.

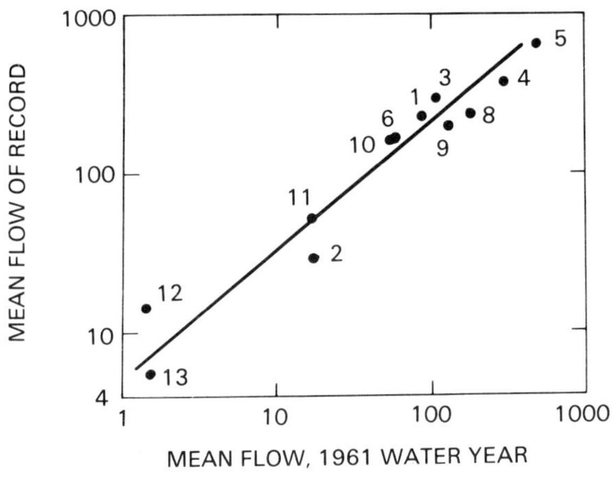

Fig. 7.6. Relation of 1961 mean flows to mean flows of record in cfs, at gaging stations in Figure 7.3 (except gage 7).

7.3.2 Low-flow characteristics from base-flow measurements

The procedure is described in Chapter 9.

7.3.3 Flow characteristics from channel size

Channel morphology studies have shown consistent relations between discharge and the corresponding width, depth, and velocity in natural channels, particularly at or near bankfull stage (Leopold, Wolman, and Miller, 1964). The geometry of the channel of a natural stream is thought to be developed by discharges near bankfull. Thus bankfull discharge should be related to channel geometry, subject of course to variations due to the types of bed and bank material in which the channel is formed and to the sediment load.

Flood characteristic discharges at various recurrence intervals have been shown to be related to channel width. A common relation is

$$Q = a W^n$$

where Q is a flood characteristic, W is channel width, and a and n are coefficients.

Channel geometry representative of flood characteristics occurs only in certain stream reaches. A straight, narrow reach is best. The bed and banks should be stable but should be of a material that has permitted the channel to develop to a size just adequate to handle the flow regimen. Most reaches suitable for slope-area measurements would be suitable for channel-geometry measurements. Proper selection of the cross section and of the level within the cross section at which the width is measured is critical to success. In meandering channels the most restricted section is just downstream from a bend. The level at which the width is measured is defined by channel features. Two levels are widely used; they identify the active-channel section and the whole-channel section (Fig. 7.7). The width of the active-channel section is a within-channel dimension represented by (1) the width of the low-water channel, (2) the distance between within-channel bars, or (3) the distance between annual vegetation lines.

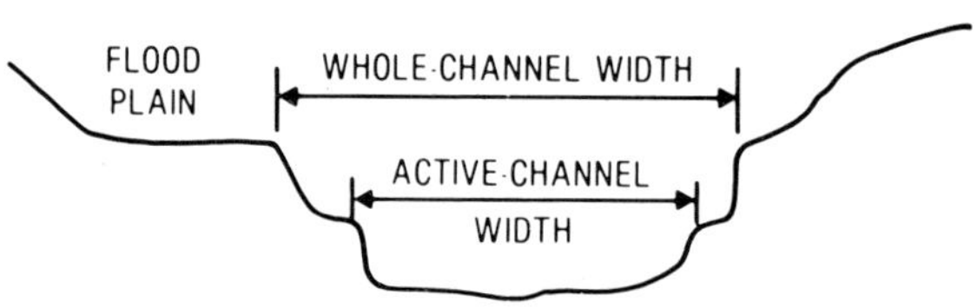

Fig. 7.7. Idealized stream cross section showing two width measurements.

The reference level for the whole-channel section is variously defined by breaks in bank slope, by the edges of the flood plain or by the lower limits of permanent vegetation. In perennial streams the whole-channel width is the width at bankfull stage. In ephemeral streams a flood plain may not exist. Width measurements are shown in Figures 7.8 and 7.9.

To define a calibration, flood characteristics from gaging-station or crest-stage-gage records are plotted on log paper against the corresponding channel widths measured near the gaged sites as in Figure 7.10. Relations of 10-year flood to width generally differ for perennial and for ephemeral streams because of the differing channel shapes. And even among perennial streams the channel shapes differ according to the material in which the channels have developed. An index of channel shape such as some function of depth may appreciably improve the calibration if a wide range of shapes are included in the sample. Channel shape is also a function of bed and bank material, and indices of these may be used (Osterkamp, 1977).

Channel geometry has also been used for estimating mean streamflow (Hedman, 1970). A relation between channel size and mean flow presumably exists because

Fig. 7.8. Whole-channel width of a perennial stream is shown by the tape.

Fig. 7.9. Measuring active-channel width of an ephemeral stream.

of a relation of mean flow to flood characteristics; such a relation for perennial streams obviously would not hold for ephemeral streams. Thus, the geographical area of applicability of a relation for estimating mean flow from channel width would be much more limited than one for estimating flood-peak characteristics.

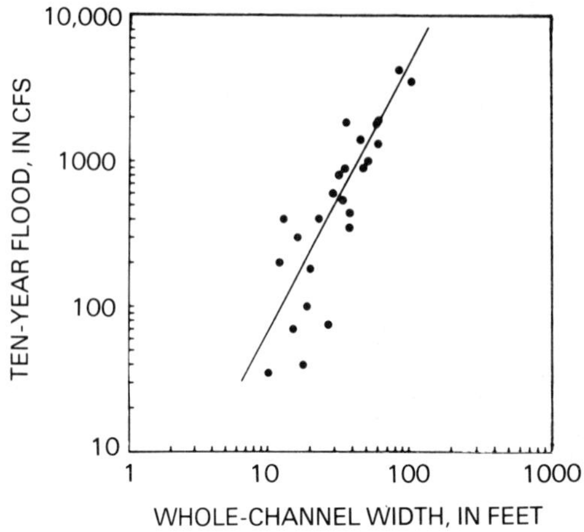

Fig. 7.10. Relation of 10-year flood to channel width for streams in Owyhee County, Idaho (From Riggs and Harenberg, 1976).

The reliability of a flood estimate made by the channel-geometry method at an ungaged site depends principally on the standard error of the calibration and on the reliability of the width measurement at the site. Wahl (1977) described a test to determine the reliability of measurements of channel width; seven individuals independently visited 22 sites and measured channel dimensions for 3 different reference levels in sections of their own choosing. Wahl attributed an average standard error for discharge of about 30 percent to differences in width measurements alone. Combining that with an assumed 40 percent calibration standard error (lower than most) gives a 50 percent standard error of the discharge estimate.

Estimating relations applicable to many regions of western United States have been published, for example see Hedman and Osterkamp (1982). In general these relations are for regions in which flow characteristics are not closely related to drainage area. The channel-geometry method is applicable to most streams but its use is recommended only where basin characteristics are poor indicators. Better estimates of flood characteristics usually can be obtained from basin characteristics in humid regions. A disadvantage of the channel-geometry method is the need for a channel measurement near the site where the flow characteristic is needed. See section 8.10.3 for a way of reducing the field work required to apply the technique.

Applications and limitations of the channel-geometry method and references to published reports are given by Riggs (1978). See also Wahl (1984).

REFERENCES

Hedman, E.R., 1970, Mean annual runoff as related to channel geometry in selected streams in California: U.S. Geol. Survey Water-Supply Paper 1999-E.

Hedman, E.R. and Osterkamp, W.R., 1982, Streamflow characteristics related to channel geometry of streams in western United States: U.S. Geol. Survey Water-Supply Paper 2193, 17 p.

Leopold, L.B., Wolman, M.G., and Miller, J.P., 1964, Fluvial processes in geomorphology: San Francisco, W.H. Freeman & Co., 522 p.

Osterkamp, W.R., 1977, Effect of channel sediment on width-discharge relations, with emphasis on streams in Kansas: Kansas Water Resources Board Bull. No. 21, 25 p.

Riggs, H.C., 1969, Mean streamflow from discharge measurements: Bull. of International Assoc. of Scientific Hydrology, Vol. XIV, No. 4, p. 95-110.

Riggs, H.C., 1978, Streamflow characteristics from channel size: Journal of Hydraulics Division, ASCE, Vol. 104, No. HY1, January 1978, p. 87-96.

Riggs, H.C. and Moore, D.O., 1965, A method of estimating mean runoff from ungaged basins in mountainous regions: U.S. Geol. Survey Prof. Paper 525-D, p. 199-202.

Thomas, D.M. and Benson, M.A., 1970, Generalization of streamflow characteristics from basin characteristics: U.S. Geol. Survey Water-Supply Paper 1975, 55 p.

Wahl, K.L., 1977, Accuracy of channel measurements and the implications on estimating streamflow characteristics: Jour. Research, U.S. Geol. Survey, Vol. 5, No. 6, p. 811-814.

Wahl, K.L., 1984, Evaluation of the use of channel cross section properties for estimating streamflow characteristics: U.S. Geol. Survey Water-Supply Paper 2262, p. 53-66.

Chapter 8

FLOOD-FREQUENCY ANALYSES

8.1 INTRODUCTION

The principles given in sections 4, 5, and 7 for defining the magnitude-frequency relations of streamflow characteristics, both gaged and ungaged, were illustrated by applications to floods. But the value of reliable flood information and the range of problems associated with getting that information justify a more detailed discussion.

8.2 ANNUAL FLOODS

An annual flood is the highest peak discharge during a year. Most flood-frequency analyses are concerned with estimating the characteristics of annual floods. Individuals that make up a population should be independent of each other and should occur randomly if a sample from that population is to provide a reasonable estimate of the frequency characteristics. Annual floods usually meet the first requirement, but short records of floods tend to be made up of nonrandom events because the causative major storms are not randomly distributed in time.

8.3 FLOODS ABOVE A BASE

An objection to the use of the annual flood series is that the second highest flood in some years is higher than the annual floods for other years but it is not considered in defining the frequency curve. This objection can be resolved by using all independent floods above an arbitrary base discharge during the period of record. These floods are called the partial-duration series, the partial-flood series, or floods above threshold. The base discharge is selected so that several peaks will qualify in most years. Only those peaks that are reasonably independent of each other should be included; independence is assumed if the peaks are widely separated in time and by a substantial recession in discharge.

8.4 ANNUAL AND PARTIAL-DURATION FREQUENCY CURVES

Frequency curves based on the annual series and on the partial-duration series are similar above a recurrence interval of about 10 years (Fig. 8.1). The theoretical relation between the two frequency curves was demonstrated by Langbein (1949). Empirical relations approximate the theoretical one but indicate some geographic variation.

The frequency curve of annual floods is used for defining the probabilities of floods that are exceeded infrequently. For probabilities of less than 0.10

(recurrence intervals greater than 10 years) either the annual or the partial-duration frequency curve could be used but annual floods are more readily available than floods above a base so the annual curve is more widely used.

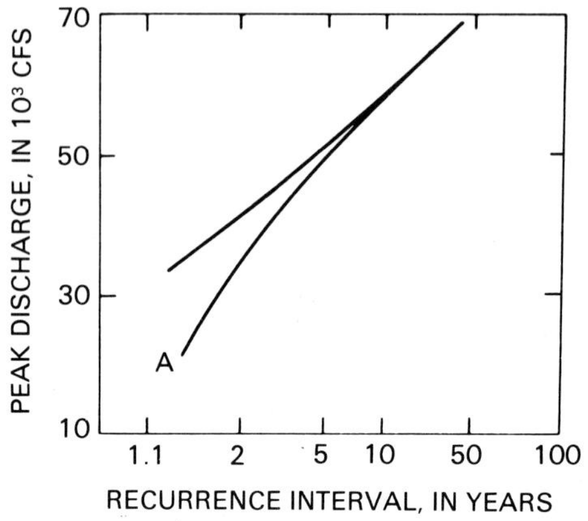

Fig. 8.1. Annual and partial-duration flood-frequency curves for Clark Fork near St. Regis, Montana.

The partial-duration frequency curve is used for evaluating risk for a period of a few years or where all exceedances are of interest, not just the largest one of the year.

8.5 FITTING ANNUAL FREQUENCY CURVES

Graphical frequency analysis is described in section 4. Interpretation of the plotted points by drawing a mean line is somewhat subjective, particularly at the high recurrence intervals. The Weibull plotting formula,

$$RI = (n+1)/m,$$

commonly used for the annual series also has been used for the partial-duration series. The number of years of record, n, is the same for both series. In the annual series the order number, m, ranges from one to n. But in the partial-duration series, m will take values much greater than n; thus the formula produces some recurrence intervals less than one which cannot be interpreted as the reciprocal of probability. The abscissa scale for a partial-duration frequency curve should be probability of exceedance or number of floods per 100 years exceeding a given value.

Mathematical fitting of flood-frequency curves eliminates the subjectivity of graphical fitting although the selection of a suitable distribution remains subjective. The most common theoretical distributions now used for annual floods are log Pearson Type 3, log normal, and the extreme-value (Gumbel). The log-normal distribution is the same as the log Pearson Type 3 with zero skew. Methods of fitting are given in section 4. Digital computer programs are commonly used.

Mathematical methods of defining flood-frequency characteristics are not suitable without modification if some of the annual floods are unusually small. In Figure 8.2 the one extremely small annual flood introduces so much curvature (negative skew) into the fitted curve that the upward extrapolation of that curve is unrealistic. Hydrologically the occurrence of an unusually low annual flood should be unrelated to the flood potential of the stream, and consequently it should not influence the upper end of the flood distribution. Such reasoning is the basis for modifying the purely statistical fitting process.

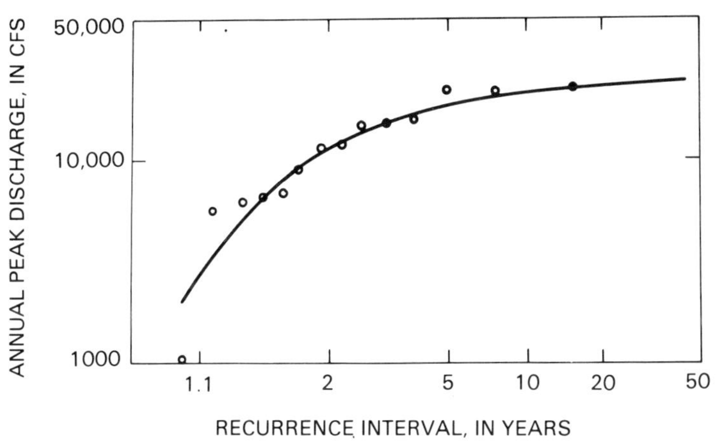

Fig. 8.2 Frequency curve for Platte River, Missouri, showing how a low outlier distorts the upper end of the frequency curve.

Deletion of the unusually low annual flood or floods before fitting the distribution mathematically is a common practice. If the lowest peak discharge in Figure 8.2 had not been used in the curve fitting, the 50-year flood would have been 40 percent greater than that shown. This is still well below the 50-year floods obtained by a graphical interpretation or by a Gumbel fit, both of which are confirmed by the subsequent record which includes 2 peaks over 50,000 cfs.

The flexibility of the log Pearson Type 3 distribution is sometimes a disadvantage; for the record in Figure 8.2 it gives too much weight to peaks on both ends of the array. Objective methods for identifying low "outliers" and for eliminating them from the fitting process are given by Water Resources Council (1981).

Some annual flood records of small, ephemeral streams include zero flows for several years. In other records the annual peaks are available only for those years in which the peak was above some base; such a record might be obtained from a crest-stage gage which was not low enough to record the stages of all annual peaks. A record containing zeros is that of Tusler Creek tributary, Montana, for 1957-73; the annual peaks are 1.71, 0, 1.04, 2.38, 0, 0, 0.84, 45.2, 0.03, 0, 0, 0, 0.08, 1.12, 1.00, 4.20, and 0.17 m^3/s. The frequency curve can be defined graphically from the 8 highest discharges in 17 years; the rest of the annual peaks need not be plotted (Fig. 8.3).

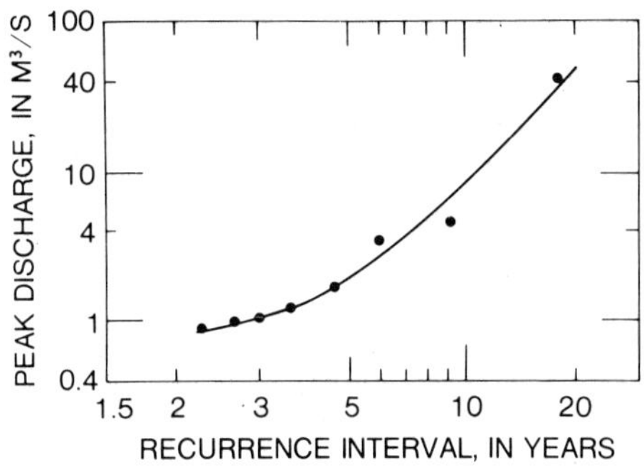

Fig. 8.3. Graphically-defined flood frequency curve for Tusler Creek Tributary, Montana, 1957-73.

Frequency curves can be defined mathematically from records containing some annual floods whose magnitudes are only known to be less than some base. The frequency curve for Tusler could be computed by fitting the log Pearson Type 3 distribution to the 8 highest discharges and then adjusting that distribution to account for the additional years of record (Jennings and Benson, 1969). An example of this type of computation is given by Water Resources Council (1981, p. 12-32).

Flood information at a gaged site may include stages of major floods that occurred prior to the gaged period. Estimates of discharges of these historic floods may be incorporated in a graphical frequency analysis by assigning appropriate recurrence intervals. For example, given a flood-peak record from 1940 to 1980 and an estimate of the discharge of the 1916 flood which is known to be larger than any subsequent one at the site, the 1916 estimate would be plotted at 65 years on the graph. Dalrymple (1960) describes how other combinations of histories and within-record floods can be analyzed graphically. The guidelines recommended by the Water Resources Council (1981) for determining flood frequency curves mathematically include procedures for incorporating historic flood information.

8.6 A UNIFORM METHOD

Flood characteristics are used for various purposes and by various people. Conflicts or inconsistencies may arise if different flood frequency curves are derived from the same data. Uniform methods of analysis reduce such problems.

Bulletin No. 15, "A uniform technique for determining flood flow frequencies", issued by the Hydrology Committee of the United States Water Resources Council in 1967, recommended use of the log-Pearson Type 3 distribution as the base method but allowed for other methods of analysis if justification were provided. Guidelines in subsequent Bulletins in 1976, 1979, and 1981 specify use of the log Pearson Type 3 distribution and provide specific procedures for using historic data, for identifying and adjusting for the effects of high and low "outliers", and for weighting skew coefficients computed from a flood record with generalized mapped skew coefficients (Water Resources Council, 1981).

The weighting of the skew coefficient computed from the sample with an average skew coefficient for the geographic area was recommended because a skew coefficient computed from a sample of a few tens of years is known to have a large error. This procedure is based on the assumption that the skewness varies geographically. The adjustments for low and high "outliers" are also intended to make the sample more representative.

The decision to use the log-Pearson Type 3 distribution for flood peaks was based on comparisons of frequency curves fitted by various statistical distributions to the graphically-defined frequency curves. Flood records for many streams encompassing a wide range of sizes and characteristics were used in the tests.

The same method of testing also was used on floods of streams in Great Britain. That investigation plus other statistical considerations led to the conclusion that the Gumbel extreme-value distribution was appropriate for the majority of British streams although the conclusion was qualified by "Statistics do not point to a single distribution being certainly the correct one and therefore a choice must involve an element of subjectivity." (National Environment Research Council (1975, p. 241, 155-160).

The guidelines of the Water Resources Council for computation of flood-frequency curves at gaging stations are not widely supported by the profession. Some hydrologists question the validity of a map of generalized skew and consequently its use in modifying the frequency curve at a gaged site. But the principal objection is that no single nonsubjective statistical method will produce the best estimates under all conditions. Support for that objection is provided by Pattison (1977), commenting on Australian experience:

"The purpose of the previous editions of Australian Rainfall and Runoff was to provide an unofficial code of recommended practice in the selection and application of methods of design for various types of hydrologic problems. The work was taken by many practicing engineers, and often by the Courts, as a guide to standard practice."

"It is the view of the National Committee on Hydrology that the role of Australian Rainfall and Runoff (1958 edition) as a code of practice has been abused. There are, for example, instances of the hydrologic design of important facilities in which the requirements of good engineering practice have been ignored, to some extent, when relevant assumptions have been drawn uncritically from guidelines set out in Australian Rainfall and Runoff. Also it is apparent that some engineering and legal bodies view any hydrologic design procedure which departs from that recommended in Australian Rainfall and Runoff as being unacceptable. This is the approach traditionally taken with codes of practice, and tends to inhibit innovative design based on local knowledge, customs, and established criteria. Insistence that hydrologic design for all parts of Australia follow a pattern using a methodology prescribed in a single compendium will, in many instances, lead to designs which are technologically inferior and economically unsound."

8.7 RECORD EXTENSION

Annual floods outside the period of record may be estimated from a longer flood record or from precipitation. Except for stations on the same stream, the first technique is rarely useful for improving the frequency curve; flood peaks are so highly influenced by areal variations in precipitation and by topography that peaks from nearby streams are not closely correlated. See section 5.4.

Rainfall-runoff models may be used to synthesize long records of flood peaks, or volumes, from rainfall. Continuous-flow models such as the Stanford Watershed Model (Crawford and Linsley, 1966) or ones that model only the storm runoff (Dawdy and others, 1972) are examples. For small streams, rainfall and discharge at 15-minute or shorter intervals are needed for adequate calibration. The only other input to the model developed in Dawdy and others (1972) is daily pan evaporation. Ordinarily 8 to 10 years of record are needed to define an adequate calibration; shorter records usually include only small floods. Figure 8.4 shows the results of calibrating on 16 flood events in 1936-46 on Beetree Creek, North Carolina, a 5.41 square-mile basin. An adequate calibration usually requires that a rain gage be in the basin. If basin rainfall is not uniform, more than one rain gage will be needed.

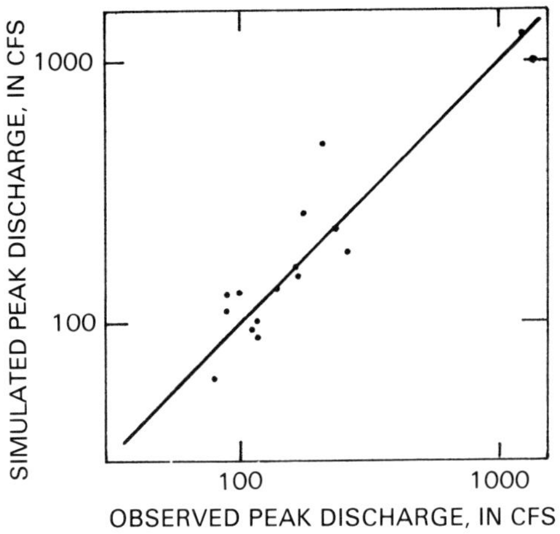

Fig. 8.4. Results of calibrating a rainfall-runoff model (After Dawdy and others, 1972).

The flood record is extended by applying a long recording rainfall record to the calibration and selecting the annual floods from the synthesized peaks. Whether the frequency curve based on the extended record is an improvement over the one based on the gage record depends on the quality of the calibration and on the representativeness of the long rainfall record as well as on the representativeness of the original gage record. It has been observed that some rainfall records of 60 or more years do not contain storms as large as others which have been obtained in the vicinity in much shorter periods. An additional uncertainty is introduced when the long rainfall record is applied to a basin distant from where the rainfall was observed.

8.8 RELATION TO BASIN CHARACTERISTICS

Flood-frequency characteristics depend on basin and climatic characteristics. If the geology and topography of a basin are more or less homogeneous and major storms cover large areas, then the flood-frequency curve can be represented adequately by a log-Pearson Type 3 distribution with a small skew. But suppose the basin encompasses a large range in elevation so that floods may result either from snowmelt or rainfall as in the Merced River basin, California. The Merced River frequency curve (Fig. 8.5) has a shape that cannot be approximated by any 3-parameter distribution. The rainfall peaks and the snowmelt peaks of Merced River are separated by several months in most years. Thus data are available for defining separate frequency curves of annual rainfall and of snowmelt peaks. These curves, shown in Figure 2.16 (Chapter 2) could have been fitted to log Pearson Type 3 distributions and the two distributions could have been combined on the basis of probabilities to get the frequency curve of annual floods. The alternative is a graphical fit to the annual floods.

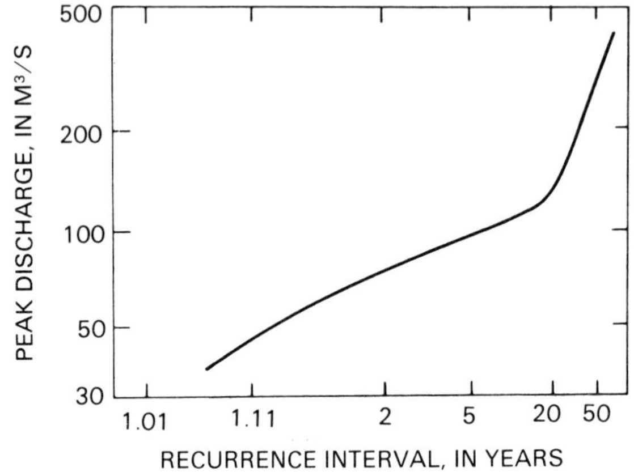

Fig. 8.5. Frequency curve of annual floods for Merced River, California (After Crippen, 1978).

More commonly, some of the flood peaks from basins of large relief are not entirely from rainfall or from snowmelt and it is not feasible to develop separate frequency curves. Nor has it proved feasible to define separate frequency curves for floods caused by frontal storms and by hurricanes because there are too few of the latter to define a frequency curve at a site and because the classification of some of the causative storms must be arbitrary. Under these conditions only a single frequency curve can be defined. Whether a single mathematically-fitted curve is reasonable can be judged from a computer plot of the curve and of the data points. An understanding of the basin hydrology and of the flood record will help in deciding whether the curve is to be accepted or modified, and if so, how.

The slopes and shapes of flood-frequency curves may change from point to point along the same stream. In regions such as the eastern slopes of the Colorado Rocky Mountains, floods in the headwaters are from snowmelt, those in the middle reaches may be from rain as well as snowmelt, and those near the Plains, particularly the major floods, are from thunderstorm rainfall. If such a stream is gaged at several points, the standard deviation and the skew of the annual floods will increase downstream.

8.9 RELIABILITY OF FLOOD-FREQUENCY CURVES

The nonrandomness of weather conditions favorable to major floods indicates that a long flood record should provide the best estimate of the frequency curve. Reliability of the larger annual discharges, if defined by indirect measurement, will also affect the frequency curve reliability but only moderately unless the discharges are greatly in error. The lack of randomness in the sample is the major unknown. For this reason, statistical confidence limits about a frequency curve are not necessarily dependable indicators of the reliability of a frequency curve, particularly if the record is short.

The upper end of a flood-frequency curve may be evaluated by comparing it to the maximum floods of record in the region. If a frequency curve shows a 100-year flood in excess of anything experienced in the region, and the frequency curve is not supported by data near that level, then the upper end of the curve is questionable. Crippen and Bue (1977) compiled maximum floods in the United States and plotted envelope curves of peak discharge versus drainage area for each of 17 regions of the conterminous United States. Their nationwide envelope curve is in Figure 8.6.

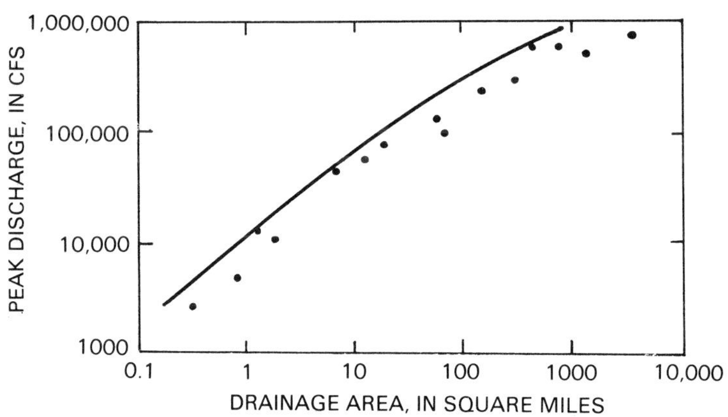

Fig. 8.6. Envelope curve of maximum floods in conterminous United States (After Crippen and Bue, 1977).

Under certain conditions historic flood stages can be estimated from geological and botanical evidence. See Stewart and Bodhaine (1961), Helley and LaMarche (1973), Costa (1978), and Baker and Kochel (1979). These historic stages, when translated to discharge, may permit realistic extrapolations of frequency curves based on short records. On the other hand, Jahns (1947) concluded "Because the Connecticut River has been engaged in lowering its bottom since the draining of the late glacial valley lake, discussion of so-called

1000-year floods on the basis of their crest heights as referred to points along the valley is of no value."

Extrapolations of frequency curves to recurrence intervals greater than 100 years generally are not reliable. Most frequency curves, graphically or mathematically derived, are unbounded at the upper end although there must be some physical limit to the flood potential on any stream. Such a physical limit has not been shown by flood records, either because they are too short or because the records are nonhomogeneous due to changes in the basins although an upper level seems indicated by estimates of major floods on Han River, China, from stage information; Chen Chia-chi and others from the Ministry of Water Conservancy and Electric Power in Peking reported in 1974 that the maximum flood on Han River occurred in 1583 and apparently was the highest in the last 1000 years.

Another approach to defining the probable maximum flood is through estimates of the maximum probable precipitation on the basin. See section 11.2.7.

Evaluation of frequency curves in the range below 100-year recurrence interval may be done subjectively by comparison among several in a region. If flood records are available at several sites on the same stream a plot of peak discharges for several recurrence intervals against channel distance, or drainage area, may be used to identify inconsistencies. Figure 8.7 shows questionable information at one site.

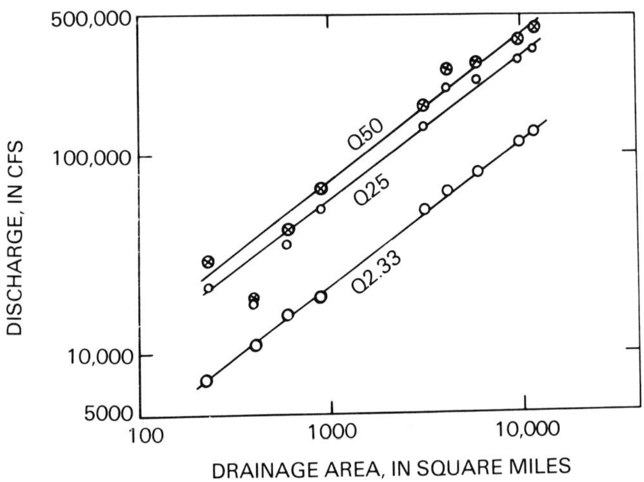

Fig. 8.7. Flood characteristics at gaging stations on Potomac River.

8.10 FLOOD CHARACTERISTICS AT UNGAGED SITES

8.10.1 Regression on basin characteristics

Refer to Chapter 4 for description of regression analysis and to Chapter 7 for its use in estimating flow characteristics of ungaged streams. The method is widely used to develop equations for estimating flood-frequency characteristics from basin characteristics. Applications of the derived relations to ungaged sites usually require only site data that are readily available from maps or weather records. The equations developed by Thomas and Corley (1977) for Oklahoma streams are typical:

$Q_2 = 0.111 \; A^{0.66} \; S^{0.23} \; P^{1.92}$		SE = 48%
$Q_5 = 1.00 \; A^{0.67} \; S^{0.26} \; P^{1.45}$		40
$Q_{10} = 2.99 \; A^{0.68} \; S^{0.28} \; P^{1.22}$		39
$Q_{25} = 9.49 \; A^{0.69} \; S^{0.30} \; P^{0.97}$		40%
$Q_{50} = 20.0 \; A^{0.69} \; S^{0.31} \; P^{0.81}$		42
$Q_{100} = 38.6 \; A^{0.70} \; S^{0.32} \; P^{0.67}$		45

where Q is the annual flood peak in cfs at the indicated recurrence interval, A is drainage area in square miles, S is an index of main-channel slope, and P is mean annual basin precipitation in inches. The standard error of each regression is given. The above equations are transformed from the log-linear form in which they were solved by multiple regression.

$$\log Q = \log a + b_1 \log A + b_2 \log S + b_3 \log P$$

Various other basin characteristics have been used in regressions for estimating flood-peak characteristics. The appropriate characteristics depend on the hydrology of the region and the availability of data. Rarely are more than three basin characteristics both statistically and practically significant. Although a fourth characteristic may be statistically significant and its inclusion may reduce the standard error a few percent, the results obtained by the equations with and without this characteristics usually do not differ appreciably.

A regression on basin characteristics is defined by data from some region, and the results are considered applicable to that region. The applicability to all parts of the region can be appraised subjectively by plotting on a map at each data site the ratio of the discharge by regression equation to the known discharge. The geographic distribution of these residuals should be more or less random.

The above test is limited to sites used in defining the relation. If these sites do not represent the range of hydrologic conditions in the region then the relation may not apply to some ungaged streams in that region. A regional relation is commonly qualified as to its applicability.

8.10.2 Index-flood method

The index-flood method was used for most of the regional flood-frequency analyses made by the U.S. Geological Survey prior to 1965 and is used in England (National Environment Research Council, 1975). It consists of two parts.

The first part graphically relates mean annual flood to drainage area, and sometimes to other variables. The mean annual flood was defined as the 2.33 year recurrence interval flood, based on the Gumbel distribution. Usually the plotted points define several different relations. On the basis of these preliminary relations, the geographic area being studied is divided into subareas such that a single relation of mean annual flood to drainage area applies to each. Thus the regionalization of the mean annual flood is attained.

The second part of the regionalization process averages the individual frequency curves to provide a regional curve. This is accomplished after expressing the flood magnitudes at selected recurrence intervals for each curve as ratios to the mean annual flood (the index flood). If some of the dimensionless individual curves are greatly different from others, the geographic area is subdivided so that each subdivision contains curves of similar shape. Then the curves in each subdivision are averaged. The subdivisions for this purpose are usually not coincident with the subareas defining the various relationships of mean annual flood to drainage area. Figure 8.8 shows the two parts of the analysis.

The index-flood method thus accomplishes the general purposes of a regionalization by relating the position of the frequency curve on the discharge scale to basin size, and by averaging the shapes of the individual curves. The method provides satisfactory results in many regions and is fairly simple to perform. The results are easy to apply to ungaged areas because usually only drainage area need be measured.

Application of the method requires arbitrary decisions as to the boundaries of subareas considered homogeneous with respect to mean annual flood or to shape of frequency curve. No subarea should be represented by fewer frequency curves than needed to define a meaningful regionalization, even though a close agreement among frequency characteristics in the subarea is not attained.

An evaluation of the index-flood method is described by Benson (1962). A detailed discussion of the preparation of regional frequency curves is given by National Environment Research Council (1975, p. 170-185).

8.10.3 From channel geometry

In arid and semiarid regions and in ones with heterogeneous geology, channel geometry is a better indicator of flood-peak characteristics than basin

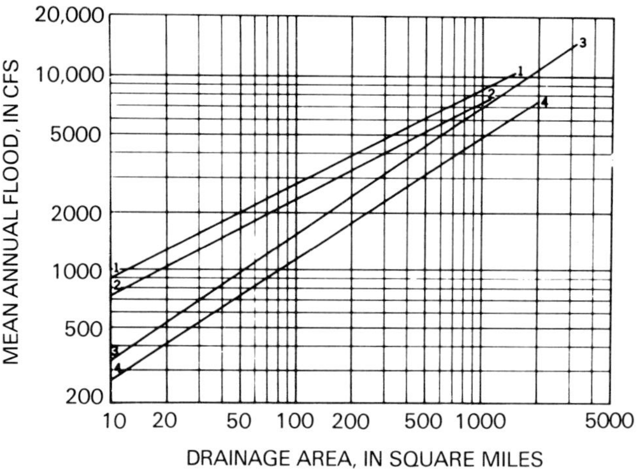

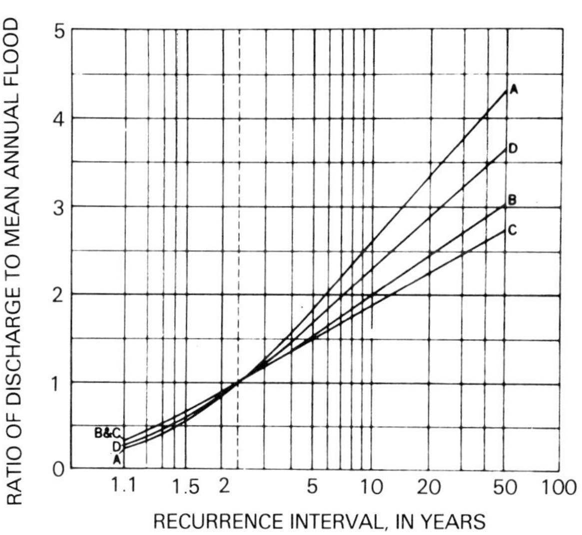

Fig. 8.8. Relations developed by the index-flood method (After Barnes and Golden, 1966).

characteristics. The method, described in Chapter 7, requires field measurements of channels for calibration and for application to ungaged sites; these measurements should be made by people with some experience with the technique. Primary calibrations are usually made with the 10-year flood because that flood is defined at many more sites than floods of higher recurrence interval. In regions where flood records are long, usually for the larger streams, other flood characteristics may be related to channel width. Figure 8.9 relating 50-year flood to channel width, is defined by data from mountain streams in Colorado, Nevada, Northwestern United States, and Alaska.

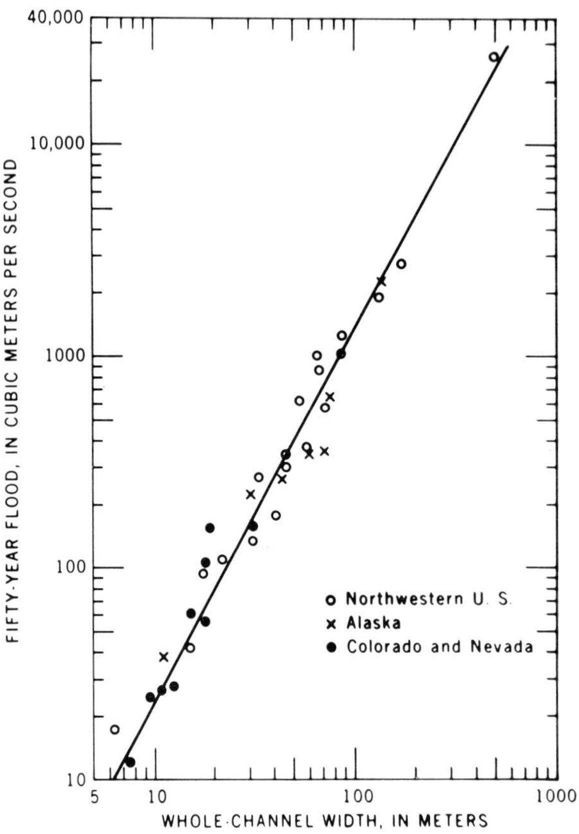

Fig. 8.9. Relation of 50-year flood to channel width (From Riggs, 1978).

Variations in the relations of flood peak characteristics to width occur because of differences in channel shapes. Channel shape depends on the material in which the channel is formed, the amount of sediment carried, and on the flow regime, particularly whether it is perennial or ephemeral. Some differences among calibrations on width are shown in Figure 8.10; the Kansas channels are deep and narrow, the western mountain channels tend to be wide and shallow, and the Kentucky channels are intermediate.

Some additional measure of channel geometry is needed for calibration in a region which contains channels of various types. Using data for 42 channels of very different shapes and sizes in Kansas, Alaska, Northern Canada, and Wyoming the standard error of regression of the 10-year flood on width and mean depth

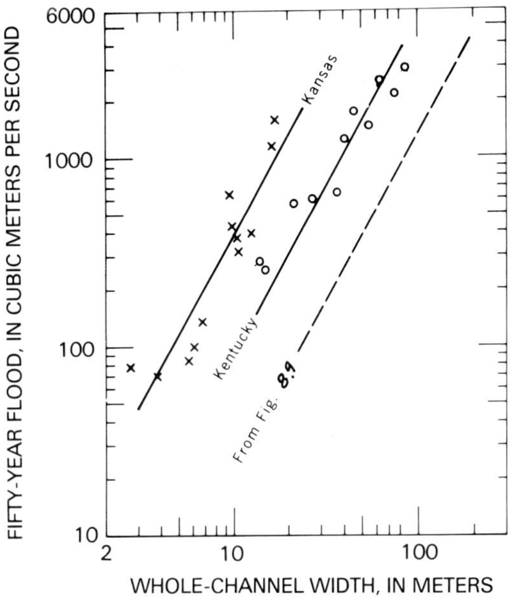

Fig. 8.10. Variations due to different channel shapes (From Riggs, 1978).

was 0.22 log units compared to 0.45 log units for a regression based on width alone. Even though mean depth was a highly significant and useful variable in this example, the standard error is unacceptably large. Calibration on streams in a smaller region usually will produce better results.

A disadvantage of the channel-geometry method is the need for field measurements at the sites of application. The user not only needs to visit each site where he wants a flood estimate but he also should have some understanding of channel morphology in order to get appropriate measurements. Riggs and Harenberg (1976) calibrated 10-year floods at gaged sites in Owyhee County, Idaho on channel width; then they measured channel widths at many ungaged sites, determined the 10-year floods at these sites, and plotted them on a map. The user can estimate the 10-year floods at other sites and on other streams by interpolation or, by using this map in conjunction with a topographic map, he can identify a measured stream similar to the one for which an estimate is wanted.

If a calibration is available only for the 10-year flood, floods of larger recurrence intervals may be approximated by multiplying the estimated 10-year flood by factors based on frequency curves for the region; this is similar to the procedure used in the index-flood method.

8.10.4 From precipitation

Methods of estimating flood-peak characteristics directly from precipitation include the so-called Rational Method and procedures developed by the U.S. Soil Conservation Service (SCS) and by the U.S. Corps of Engineers Hydrologic Engineering Center (HEC).

The Rational Method, widely used for design of storm drainage facilities, gives peak runoff rate, Q, in cubic feet per second, by

$$Q = C\ I\ A$$

where C is a dimensionless runoff coefficient; I is rainfall intensity in inches per hour for a period of time equal to the time of concentration of the basin, and for the same recurrence interval as the discharge; and A is the drainage area in acres. Time of concentration is usually defined as the estimated time required for runoff to flow from the farthest point in the drainage area.

The coefficient C depends on the permeability of the basin soils and other surfaces. Its value ranges from about 0.10 for sandy soil to 0.95 for impervious pavement; it is generally above 0.50 for urbanized areas. Rainfall intensity, I is obtained from an intensity-duration-frequency relation derived from precipitation records. The duration used is the time of concentration which must be estimated at ungaged sites. Anderson (1970) shows lag time (time of concentration) as a function of basin length divided by the square root of basin slope for natural and for fully urbanized basins.

The Rational Method usually is applied only to small drainage areas. Schaake, Geyer, and Knapp (1967) examined the method and compared runoff estimates with those from runoff frequency curves for urban drainage areas. They found that for every five Rational Method estimates made by storm drain designers in the Baltimore, Md. area, one of them, on the average, may be in error by more than 25%. Ardis, Dueker, and Lenz (1969) reported a wide variation in procedures used by practicing engineers for applying the Rational Method, and a consequent lack of consistency among results.

Both the Hydrologic Engineering Center (HEC) of the Corps of Engineers and the Soil Conservation Service (SCS) have computer programs for estimating flood peaks of various recurrence intervals from precipitation. These programs are applicable to drainage areas ranging from a few to several thousand square miles.

The SCS TR-55 graphical method of estimating flood-peak characteristics begins with classification of all soil units in the basin as to infiltration capacity. This classification is used with adjustment for soil cover to define a curve number (CN) for each. These are area weighted and the result is used with storm rainfall to get direct runoff in inches, usually by a graphical solution of a runoff equation. Peak discharge is computed from direct runoff computed as above; drainage area; time of concentration from channel length and slope; and a constant (U.S. Soil Conservation Service, 1971), and (McCuen, 1982).

Methods for estimating flood peaks from precipitation usually require the assumption that the recurrence interval of the computed flood is the same as the recurrence interval of the causative rainfall, a questionable assumption in some drainage basins. The methods also require input data that often are either unavailable or that must be defined somewhat subjectively.

8.10.5 Interpolation along channel

Streamflow records and thus flood-peak characteristics are available at intervals along the larger streams in the United States. Better estimates between the gaged sites can be made by interpolation than by any of the methods described above, particularly if the flood characteristics do not change consistently along the channel. The interpolation shown in Figure 8.11 is based on 10-year floods defined at 3 sites on the main stream and on 5 tributaries in the

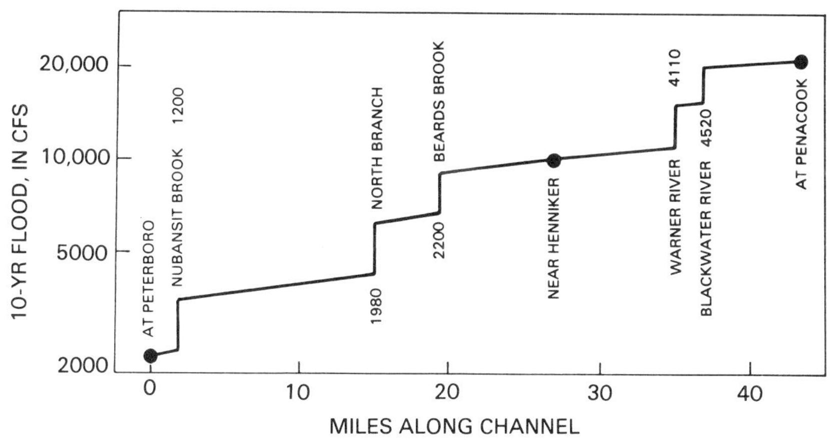

Fig. 8.11. Ten-year floods interpolated between gaged sites, Contocook River, New Hampshire. Figures are 10-year floods at tributaries.

reach. The abscissa is distance along the channel. The gaged points are plotted and the mouths of major tributaries are located. Interpolation is by trial, based on the assumption that appreciable increases occur only where tributaries enter. The contribution of a gaged tributary should be not more than its 10-year flood but it may be much less if the tributary flood is usually not concurrent with the flood on the main stream.

The plot of Figure 8.12 shows that the higher recurrence-interval floods decrease downstream on Powder River. This generally indicates either a lack of major tributaries or that the tributaries contribute relatively little to the main channel flood discharge.

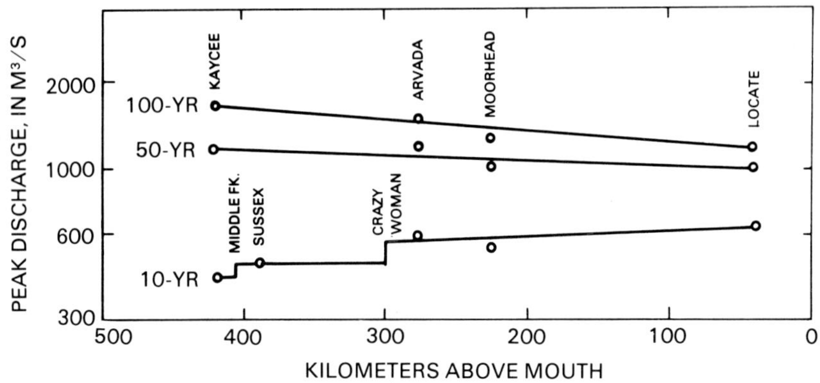

Fig. 8.12. Flood characteristics along Powder River, Wyoming-Montana.

Plots of flood characteristics against channel distance may be used to check the consistency of estimates at gaged sites as discussed in Section 8.9. In Figure 8.12, the station data indicate a large decrease from Arvada to Moorhead for which there seems to be no physical basis. More likely the difference is due to the different periods of record on which the frequency curves are based; averaging the two discharges seems reasonable.

8.11 REGULATED STREAMS

A record of annual floods during a period when flow was regulated in a consistent manner is considered suitable for interpreting on a probability basis. But, if during the period of record an additional reservoir or reservoirs were added, or the pattern of regulation was changed, then the whole record should not be used to define a frequency curve unless the changes are considered to have a minor effect on floods at the site of interest.

Models can be used to compute homogeneous records that reflect some particular level of development in a basin, usually the present condition or a proposed future condition. Natural streamflow records can be used as input. These flows are routed through storages taking into account reservoir-operating rules, evaporation losses, tributary inflows, precipitation on water surfaces, and diversions and return flows. The synthesized record can then be analyzed by conventional methods to determine the flow characteristics. See Jeffcoat, et al (1974, 1976).

See Chapter 10 for methods of estimating flood-frequency characteristics in basins modified by man.

REFERENCES

Anderson, D.G., 1970, Effects of urban development on floods in northern Virginia: U.S. Geol. Survey Water-Supply Paper 2001-C, 22 p.

Ardis, C.V., Dueker, K.J., and Lenz, A.T., 1969, Storm drainage practices of thirty-two cities: Journal of Hydraulics Division, ASCE, Vol. 95, No. HY1, p. 383-408.

Baker, V.R. and Kochel, R.C., 1969, Long-term flood-frequency analysis using geologic data: International Assoc. of Hydrological Sciences Publ. No. 128, Canberra, p. 3-9.

Barnes, H.H., Jr. and Golden, H.G., 1966, Magnitude and frequency of floods in the United States. Part 2B, south Atlantic and eastern Gulf of Mexico basins, Ogeechee River to Pearl River: U.S. Geol. Survey Water-Supply Paper 1674, 409 p.

Benson, M.A., 1962, Evaluation of methods for evaluating the occurrence of floods: U.S. Geol. Survey Water-Supply Paper 1580-A, 30 p.

Costa, J.E., 1978, Holocene stratigraphy in flood-frequency analysis: Water Resources Research, Vol. 14, No. 4, p. 626-632.

Crawford, N.H. and Linsley, R.K., 1966, Digital simulation in hydrology, Stanford Watershed Model IV: Tech. Rept. No. 39, Dept. of Civil Engineering, Stanford University, Stanford, Calif.

Crippen, J.R., 1978, Composite log Pearson Type 3 frequency-magnitude curve of annual floods: U.S. Geol. Survey Open-File Rept. 78-352, 5 p.

Crippen, J.R. and Bue, C.D., 1977, Maximum floodflows in the conterminous United States: U.S. Geol. Survey Water-Supply Paper 1887, 52 p.

Dalrymple, T., 1960, Flood-frequency analyses: U.S. Geol. Survey Water-Supply Paper 1543-A, 104 p.

Dawdy, D.R., Lichty, R.W., and Bergman, J.M., 1972, A rainfall-runoff simulation model for estimation of flood peaks for small drainage basins: U.S. Geol. Survey Prof. Paper 506-B, 28 p.

Helley, E.J. and LaMarche, V.C., 1973, Historic flood information for northern California streams from geological and botanical evidence: U.S. Geol. Survey Prof. Paper 485-E, 16 p.

Jahns, R.H., 1947, Geologic features of Connecticut Valley as related to recent floods: U.S. Geol. Survey Water-Supply Paper 996, 158 p.

Jeffcoat, H.H., Jennings, M.E., and Peterson, J.B., 1974, Lakes Marion-Moultrie stream system investigation: Part I - Model selection, calibration and error analysis: U.S. Geol. Survey Water Resources Investigations 25-74, 55 p.

Jeffcoat, H.H., Jennings, M.E., Collins, D.L., and Shearman, J.O., 1976, Lakes Marion-Moultrie stream system investigation: Part II - simulation studies: U.S. Geol. Survey Water Resources Investigations 76-11, 26 p.

Jennings, M.E. and Benson, M.A., 1969, Frequency curves for annual flood series with some zero events or incomplete data: Water Resources Research, Vol. 5, No. 1, p. 276-280.

Langbein, W.B., 1949, Annual floods and the partial-duration series: Trans. American Geophysical Union, Vol. 30.

McCuen, R.H., 1982, A guide to hydrologic analysis using SCS methods: Englewood Cliffs, N.J., Prentice-Hall, Inc., 145 p.

National Environment Research Council, 1975, Flood studies report, 5 Vols.: Institute of Hydrology, Wallingford, Oxon, England.

Pattison, A., Ed., 1977, Australian rainfall and runoff, flood analysis and design: The Institution of Engineers, Australia, Canberra, p. iii.

Riggs, H.C., 1978, Streamflow characteristics from channel size: Journal of Hydraulics Division, ASCE, Vol. 104, No. HY1, p. 87-96.

Riggs, H.C. and Harenberg, W.A., 1976, Flood characteristics of streams in Owyhee County, Idaho: U.S. Geol. Survey Water-Resources Investigations 76-88, 14 p.

Schaake, J.C., Jr., Geyer, J.C., and Knapp, J.W., 1967, Experimental examination of the rational method· Journal of Hydraulics Division, ASCE, Vol. 93, No. HY6, p. 353-370.

Stewart, J.E. and Bodhaine, G.L., 1961, Floods in the Skagit River basin, Washington: U.S. Geol. Survey Water-Supply Paper 1527, 66 p.

Thomas, W.O., Jr. and Corley, R.K., 1977, Techniques for estimating flood discharges for Oklahoma streams: U.S. Geol. Survey Water Resources Investigations 77-54, 170 p.

U.S. Soil Conservation Service, 1971, SCS National Engineering Handbook, Section 4, Hydrology, Chapters 10 and 16.

Water Resources Council, 1981, Guidelines for determining flood flow frequency: Bulletin 17A of the Hydrology Committee, Washington, D.C.

Chapter 9

LOW-FLOW CHARACTERISTICS

9.1 INTRODUCTION

The adequacy of streamflow to meet requirements for disposal of liquid wastes, and for municipal or industrial supplies, supplemental irrigation, and maintenance of suitable conditions for aquatic life is commonly evaluated in terms of low-flow characteristics (Riggs, 1980). Certain of these low-flow characteristics are useful as variables in regional draft-storage studies, as the basis for forecasting seasonal low flows, and as indicators of the amount of ground-water flow to the stream. The U.S. Environmental Protection Agency (EPA), and some States, incorporate a low-flow characteristic into their regulations on pollution control.

An annual low flow can be defined as the lowest daily mean flow in a year but it is more commonly defined as the lowest average flow for some number of consecutive days. The 7-day low flow (Hoyt, 1938) is widely used; an average for 7 days is less likely to be affected by minor disturbances upstream than is the minimum daily flow. Annual low flows for 1, 3, 7, 14, 30, 60, 90, 120, and 183 consecutive days can be readily obtained from a streamflow record by computer. The climatic year, April 1 to March 31, is used because for most of the United States it contains the entire low flow period of each year.

The dependable supply of streamflow is commonly described by frequency curves of annual or seasonal low flows, or by a point on the frequency curve, such as the 7-day discharge at 10-year recurrence interval (the 7-day 10-year low flow). The rate at which streamflow recedes in the absence of precipitation is described by the recession curve (see Chapter 5); the recession rate is useful in forecasting low flows and in understanding the relation with ground water. The reduction in discharge of a stream may be accompanied by a degradation of water quality, principally by increased temperature, by the concentration of dissolved solids by evaporation or freezing, and by the reduced aeration capability. Recession and water-quality characteristics are not discussed further in this chapter.

9.2 FREQUENCY CURVES

Frequency curves are prepared from annual low flows by methods described in Chapter 4. Typical graphically-defined frequency curves for 7-, 14-, 30-, and 60-day periods are shown in Figure 9.1. The recurrence interval scale is based on the Gumbel distribution; a normal distribution scale could have been used.

Most low-flow frequency curves for streams that do not go dry are smooth curves that are concave upward on a log-probability graph. The probability

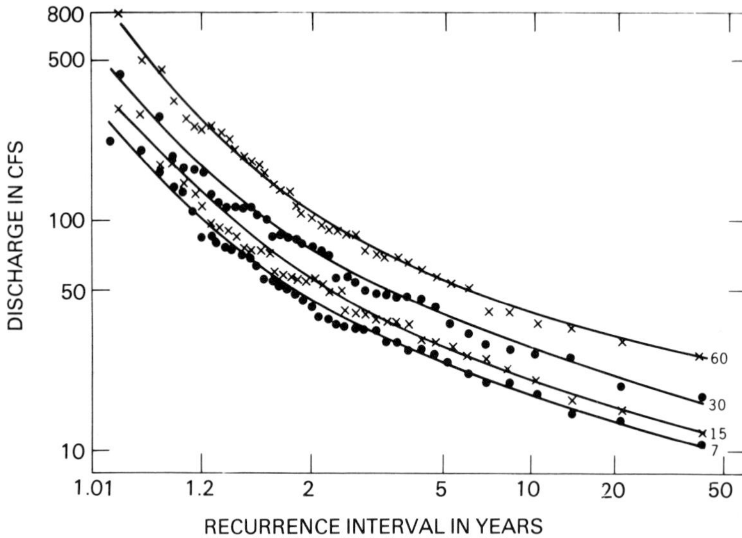

Fig. 9.1. Frequency curves of annual lowest mean discharge for indicated numbers of consecutive days, Spoon River, Illinois (From Riggs, 1980).

distribution of annual low flows at a particular site on a stream depends on the distribution of precipitation over the basin in time and space; on the temperature regimen, which may permit storage of water as snow for considerable periods and which also influences the evapotranspiration rate; and on the soil and geologic characteristics which govern the recharge and discharge rates in the basin. In addition, at a given site some of the annual low flows may be derived entirely from ground-water inflow to the stream, whereas others may include some flow from reduction of upstream storage during those years when frequent rains occur during the low-water season. Furthermore the base flow of a stream may be derived from several aquifers, not all of which contribute at all times. Consequently not all low-flow frequency curves will have a smooth, concave upward shape, nor should the high annual minimum flows at a site necessarily be considered as belonging to the same population as the smaller ones.

For examples of low-flow frequency curves having non-typical shapes see Figure 9.2. The annual minimum flow of Suwannee River at White Springs consists of (1) a flow ranging from several hundred cubic feet per second to zero from the headwaters in Okefenokee Swamp and (2) an inflow of 5 to 10 cubic feet per second from a limestone aquifer in a lower reach of the river. Curve shapes such as that of Suwannee River also result if considerable mine drainage or sewer effluent is included in the flow. The shape of the Uwharrie River curve is at least in part due to a diversion for a city supply. A heavy draft by evapotranspiration (of a more or less constant amount during the low-flow period, independent of discharge) may also help to produce the convex shape.

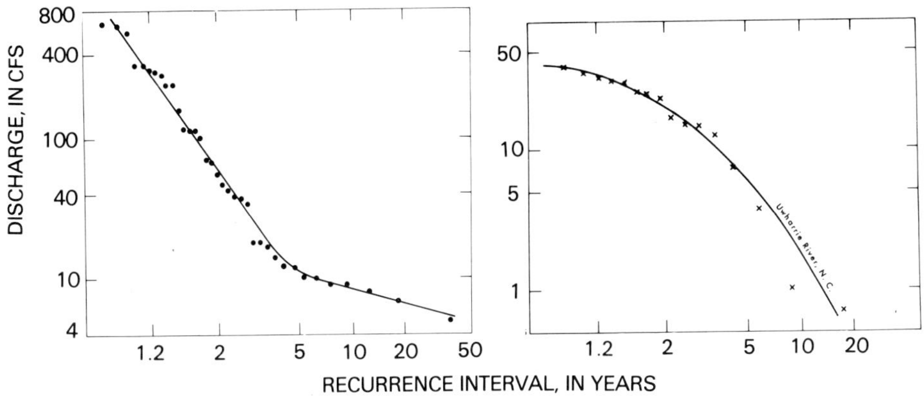

Figure 9.2. Frequency curves of annual minimum 7-day means for Suwannee River, Florida (left), and Uwharrie River, North Carolina (right).

Low-flow data such as those for Suwannee River cannot be fitted adequately at the lower end by any 2 or 3-parameter distribution. Even with data such as in Figure 9.1, a fitted log Pearson Type 3 distribution might be unrealistic at the lower end because the higher discharges affect the skew, and thus influence the shape of the lower end of the curve. A graphically-fitted curve should be considered the basic frequency curve for annual low flows, although a log Pearson Type 3 fit is adequate for most. An alternative to graphical fitting is the use of only the lower half of the array of low flows to define the lower part of the frequency curve analytically. The method of fitting partial data is given by Jennings and Benson (1969) and is described briefly in Chapter 8.

9.2.1 Interpretation

Referring to either of the previous figures, the 7-day 10-year low flow is the discharge at 10-year recurrence interval taken from a frequency curve of annual values of the lowest mean discharge for 7 consecutive days (the 7-day low flow).

The 7-day low flow will be less than the 7-day 10-year low flow at intervals averaging 10 years in length; or the probability is 1/10 that the 7-day low flow in any one year will be less than the 7-day 10-year low flow.

Comparable statements can be made for other periods of days and recurrence intervals. Note that the probability is that of being less than; it is not that of nonexceedance.

The probability interpretation, above, is strictly applicable only if the annual low flows are independent of each other. A dependence may result from a carryover of ground water from one year to the next so that the base flow is somewhat dependent on the previous years flow. Dependence may be recognized from a plot of annual minimum flows against those for the previous year. As shown in Figure 9.3, the South Fork Obion River low flows are highly correlated with those of the previous year and thus are not independent events. The Shoal Creek data appear to be uncorrelated.

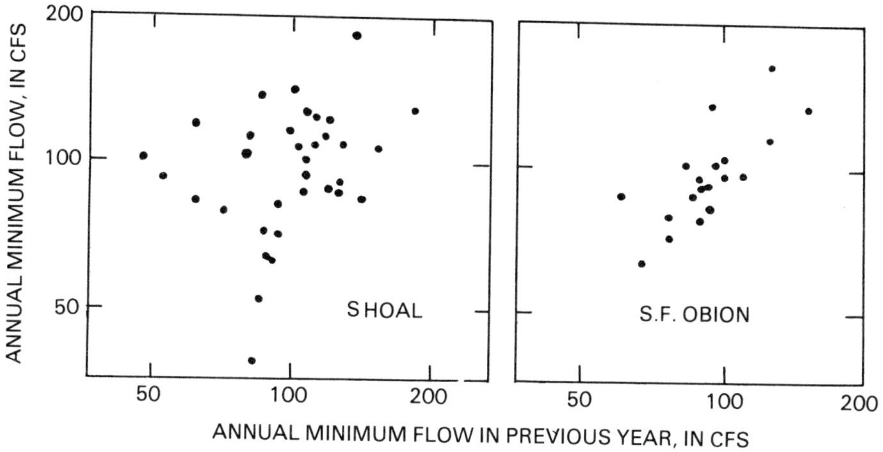

Fig. 9.3. First-order autocorrelation of low flows of two Tennessee streams.

9.2.2 Reliability

Reliability of a low-flow frequency curve usually increases with the length of record. A record of 15 or 20 years may not provide a representative sample of low flows. For example, Figure 9.4 shows frequency curves based on two 15-year periods and on the 31-year period of record. One short-period curve is higher and the other is lower than the curve based on all the record.

The tendency for drought years to occur nonrandomly suggests the desirability of providing a long time base for low-flow frequency curves. One way to do this would be to extend short records by correlation with long ones and use the combined actual and synthetic discharges to define the frequency curve. A test of this method is reported by Riggs 1972, p. 6-7). Another way would be to develop ratios of selected low-flow characteristics for various periods in time to those for the total period at one or more long-record stations. The first approach is statistically valid only when correlation with a long record is high (Fiering, 1963), and the second has not been thoroughly tested. Neither method appears to be of much practical value.

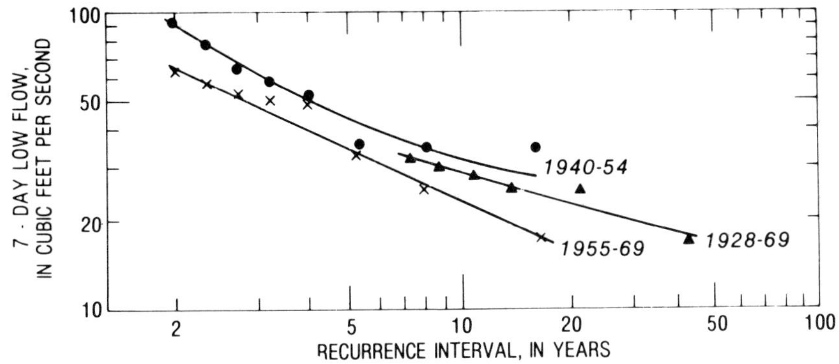

Fig. 9.4. Effect of period of record on low-flow frequency curve, Ochlockonee River, Florida (From Riggs, 1980).

Seasonal rainfall averaged over several sites may help to interpret a low-flow record. A plot of low-flow data for Tallapoosa River at Wadley, Ala., is shown in Figure 9.5 in which the 3 lowest points appear to be "out of line" with the others. A curve fitted to all the points either analytically or by eye would be concave downward. Such a shape on this plotting paper is unusual for streams with low flows of this size. Riggs (1961) related the annual minimum flows of this stream to spring and summer precipitation and found that the 3 lowest plotted points corresponded to the 3 smallest precipitation events in at least 68 years. When the recurrence intervals of the 3 points are based on the 68 year period all points on the graph can be averaged by a straight line.

9.2.3 Seasonal frequency curves

Annual low flows usually occur in the late summer or autumn in the conterminous United States although there are two low water seasons in southern Florida and some annual low flows occur in winter during freezeups in the northern states. If the concern for low flows is independent of season then the frequency curve of annual low flows is adequate. If not, seasonal, or even monthly, low flows should be used to define appropriate frequency curves. Again, only one discharge per year is used to define a particular frequency curve.

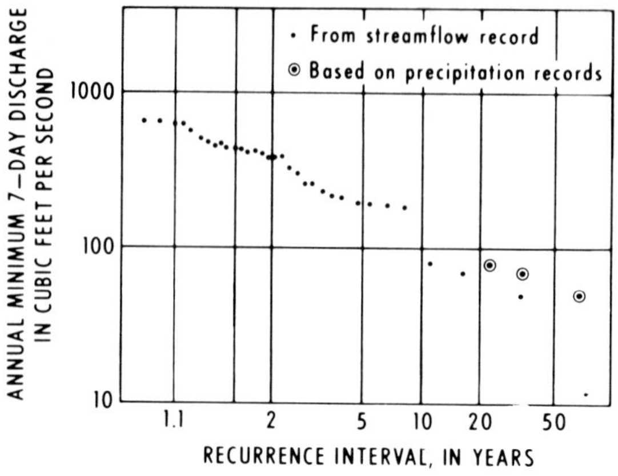

Fig. 9.5. Low-flow frequency plot showing use of precipitation records to improve estimates of recurrence interval.

9.2.4 Regulated streams

Annual minimum flows of regulated streams may be considered as random events if they are all drawn from a period in which the regulation was done in a consistent manner. And of course the frequency curve derived from those minimum flows applies only so long as that pattern of regulation continues.

Where the low flow at a site is derived from substantial upstream reservoir releases, a basin model which routes those releases downstream and includes estimates of inflows to the reach will simulate streamflow data under present conditions. These simulated data may be used to define low-flow frequency curves. Shearman and Swisshelm (1973) developed and applied such a model to the Upper Kentucky River basin.

9.3 LOW-FLOW CHARACTERISTICS AT UNGAGED SITES

The magnitudes of low base flows depend on the geology of the basin and on the losses from evapotranspiration as well as on drainage area and precipitation. Neither geology nor evapotranspiration can be reliably quantified in a basin. Consequently, regression of low-flow characteristics on basin characteristics is generally inadequate for estimating low-flow characteristics at ungaged sites. Thomas and Benson (1970) found very high standard errors for such regressions in 4 widely separated regions of the United States; their regression models did not include indices of geology or evapotranspiration.

Various methods, including regression on basin characteristics, have been used successfully in regions of limited areal extent over which the principal unmeasurable characteristics have little variation. In other regions a regression equation with a high standard error may provide an acceptable tentative low-flow estimate when an answer is needed immediately.

9.3.1 Partial record method

The most widely applicable, and the most accurate method of estimating low-flow characteristics at an ungaged site requires a few base-flow discharge measurements at the site. These measurements are related to concurrent discharges of a stream for which a low-flow frequency curve has been defined; then points on that frequency curve are transferred through the relation to get estimates at the ungaged site. Figure 9.6 shows the procedure. The base-flow measurements should be on different recessions in the season when low flows commonly occur, and preferably should be made in two or more years.

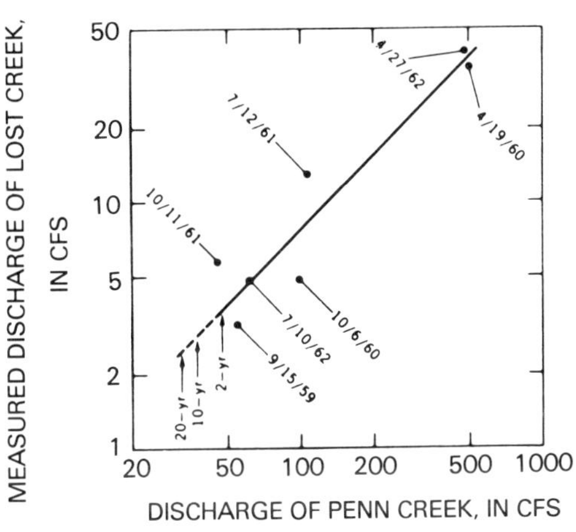

Fig. 9.6. Base-flow measurements of Lost Creek related to concurrent discharges of Penn Creek. Low-flow characteristics of Penn Creek are transferred through the relation.

Empirical tests using 24 pairs of gaging-station records indicate that the reliability of an estimate of the 7-day 10-year low flow by this method depends on how well the relation curve is defined and how far the relation needs to be extrapolated downward. In these tests one of each pair of stations was assumed to be ungaged, and low flows selected randomly from its record were assumed to represent base flow measurements. These were related to the concurrent daily mean flows at the other station. The relation line was drawn before the known 7-day 10-year low flows were plotted. Figure 9.7 is an example of a good relation requiring a small downward extrapolation. The 7-day 10-year low flow of Moshammon Creek of about 8 cfs, which would be estimated from the relation, compares favorably with 9.7 cfs based on 32 years of record. But when the concurrent base flows of two streams are not closely related and when the 10-year low flow of the index station is far below the defined relation, the estimate probably will be poor, especially if it is less than one cfs. Small discharges in large channels are subject to substantial abstractions from evapotranspiration and these abstractions vary in magnitude from year to year. The results of all 24 tests, in Figure 9.8, show that reliable estimates can usually be made for all but very low discharges.

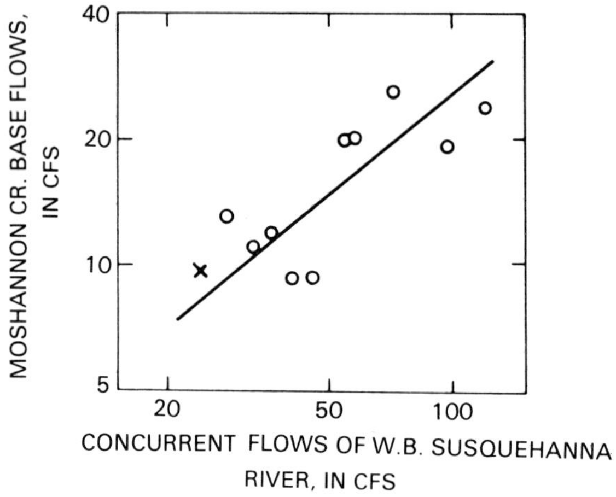

Fig. 9.7. Verification of method of estimating 7-day 10-year low flow from base-flow measurements. The X relates the known values of 7-day 10-year low flows.

Riggs (1972) recommended that the partial-record method be used to estimate the 7-day 10-year low flow even if up to 10 years of continuous streamflow record is available. Estimates based on concurrent base-flows tend to be more reliable than those based on less than 10 annual minimum flows at the site.

9.3.2 Seepage runs

Ground-water inflow to, or outflow from, some channels is concentrated in particular reaches rather than being uniform along the channel. Seepage runs will identify significant channel gains or losses and will aid in interpreting low-flow characteristics along a channel. A seepage run consists of measuring discharge at intervals along a channel reach during a period of base flow. Figure 9.9 is a plot of discharges measured on a day in late fall on a Kansas River. The tributary flows were also measured and were used to interpret the channel flows between measuring points.

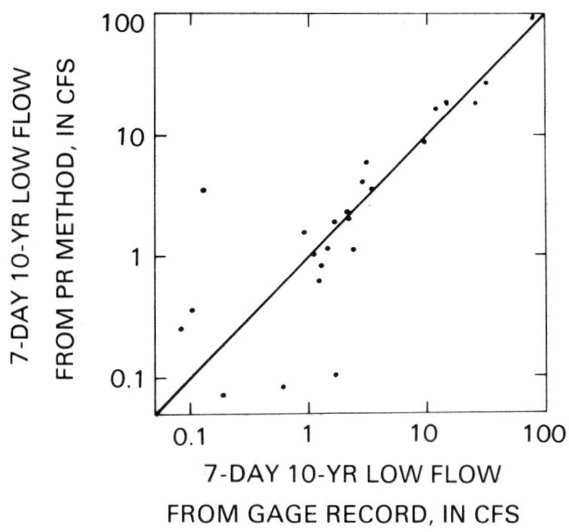

Fig. 9.8. Verification of the partial-record method based on 24 pairs of stations.

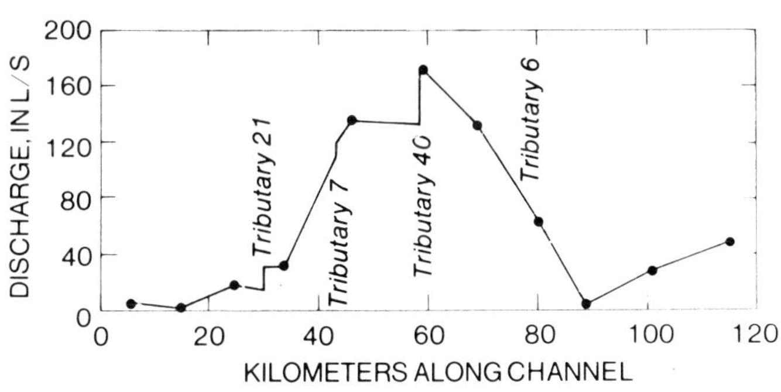

Figure 9.9. Seepage-run data and interpretation on a Kansas River. Numbers at tributaries are measured inflows.

9.3.3 Interpolation along a channel

Some extrapolation to ungaged sites may be achieved by plotting the low-flow characteristics at gaged sites against channel distance and by using low-flow characteristics at partial-record sites on tributaries to help interpret between gaged sites. Figure 9.10 was prepared from information in the report by Carter and Putnam (1977). Interpretation between gaged points may require estimating low flows of ungaged tributaries by a trial-and-error process in which the objective is to draw a line which gives reasonable and consistent increases along the channel between tributaries and gaged points.

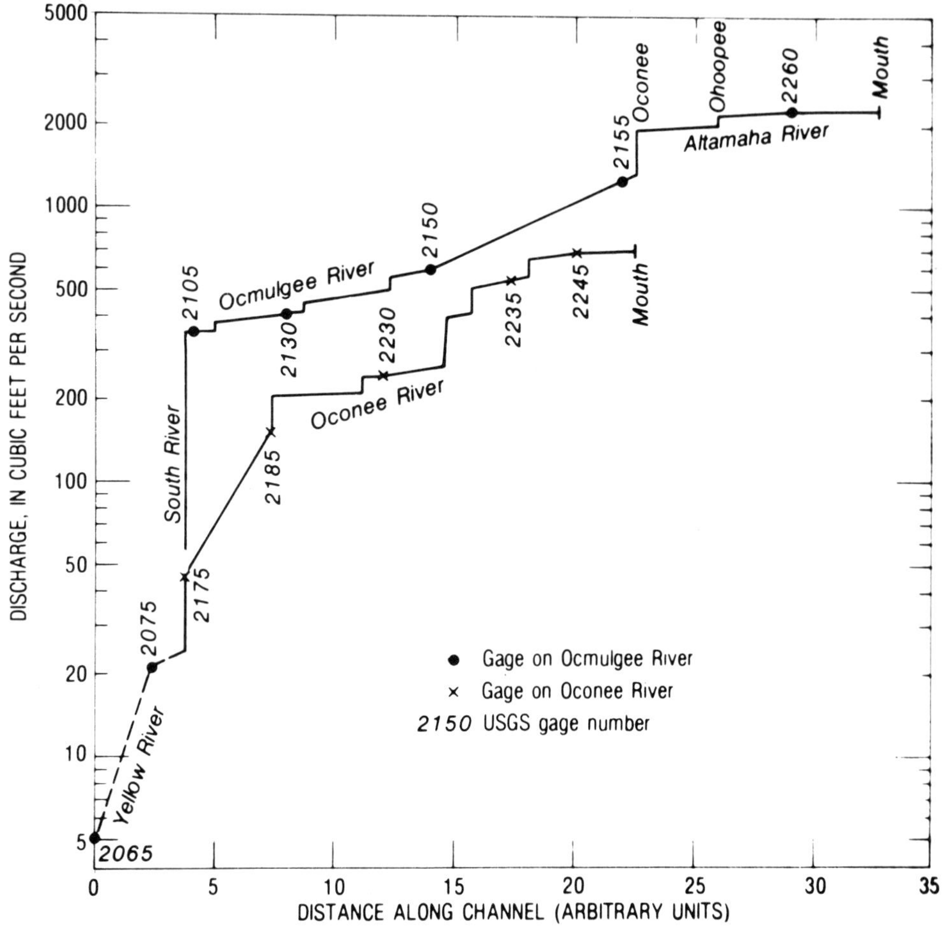

Fig. 9.10. Interpolation of 7-day 10-year low flows along channels, Altamaha River basin, Georgia (From Riggs, 1980).

If there is no significant difference between the low-flow characteristics at two gaged sites then one may assume that intervening tributaries have no flow at those times, or that the inflows and channel losses are approximately equal.

Because of the lack of information between stations 2150 and 2155 on Figure 9.10, the graph shows a uniform and quite large increase in low flow through that reach. Results of a seepage run should permit a more realistic interpretation.

REFERENCES

Carter, R.F. and Putnam, S.A., 1977, Low-flow frequency of Georgia streams: U.S. Geol. Survey Water Resources Investigations 77-127.

Fiering, M.B., 1963, Use of correlation to improve estimates of the mean and variance: U.S. Geol. Survey Prof. Paper 434 C, 9 p.

Hoyt, W.G., 1938, Drought of 1936, with discussion on the significance of drought in relation to climate: U.S. Geol. Survey Water-Supply Paper 820, p. 14.

Jennings, M.E. and Benson, M.A., 1969, Frequency curves for annual flood series with some zero events or incomplete data: Water Resources Research, Vol. 5, No. 1, p. 276-280.

Riggs, H.C., 1961, Rainfall and minimum flows along the Tallapoosa River, Alabama: U.S. Geol. Survey Prof. Paper 424-B, p. 96-98.

Riggs, H.C., 1972, Low-flow investigations; U.S. Geol. Survey Techniques of Water-Resources Investigations, Book 4, Chapter B1, 18 p.

Riggs, Henry C., Chmn., 1980, Characteristics of low flows: Jour. of Hydraulics Division, ASCE, Vol. 106, No. HY5, p. 717-731.

Shearman, J.O. and Swisshelm, R.V., Jr., 1973, Derivation of homogeneous streamflow records in the upper Kentucky River basin, southeastern Kentucky: U.S. Geol. Survey Open-File Rept., Louisville, Ky.

Thomas, D.M. and Benson, M.A., 1970, Generalization of streamflow characteristics from drainage-basin characteristics: U.S. Geol. Survey Water-Supply Paper 1975.

Chapter 10

THE CHANGING ENVIRONMENT

10.1 INTRODUCTION

Streamflow characteristics change with time in response to natural events and to man's activities. Except for long-term trends in climate, most natural events cause only short-term changes in streamflow characteristics. The effects of man's activities range from negligible to major, both in magnitude and in duration.

Knowledge of long-term climatic trends would permit better estimates of future flow characteristics to be made. But records of a hundred years or so are too short to positively identify and quantify a trend. Trends defined by periods of 30 or 40 years may be only temporary. For example, the Great Lakes annual precipitation record in Figure 10.1 shows a downward trend in the early part and an upward trend in the latter. Likewise, trends up to tens of years can be identified in some streamflow records but there is no assurance that these will continue for an appreciable time. On the other hand the flow of Labe River in Czechoslovakia (Fig. 10.2) shows no trend in over 100 years.

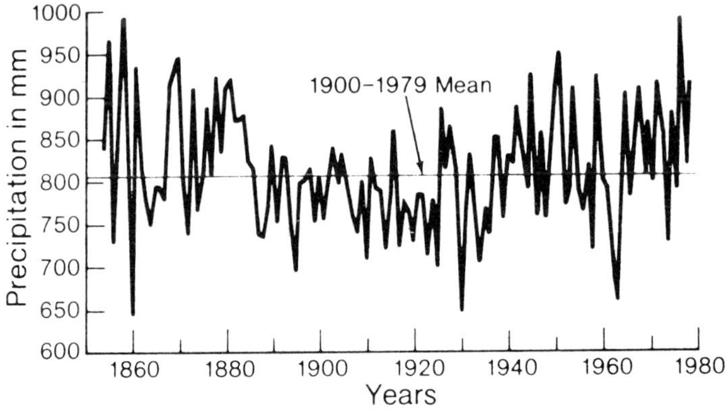

Fig. 10.1. Great Lakes annual precipitation (From Quinn, 1981).

The effects of man's activities are needed for assessing the desirability or the consequences of some proposed development. Most effects on water resources are difficult to quantify because they tend to be obscured by the natural variability of streamflow. And the effects defined on one basin may be only an approximation of that which would occur on others under the same treatment.

Generalizations are necessary, however, in planning; these can be based on analyses of basins that were changed, and on an understanding of the hydrology of the subject basins.

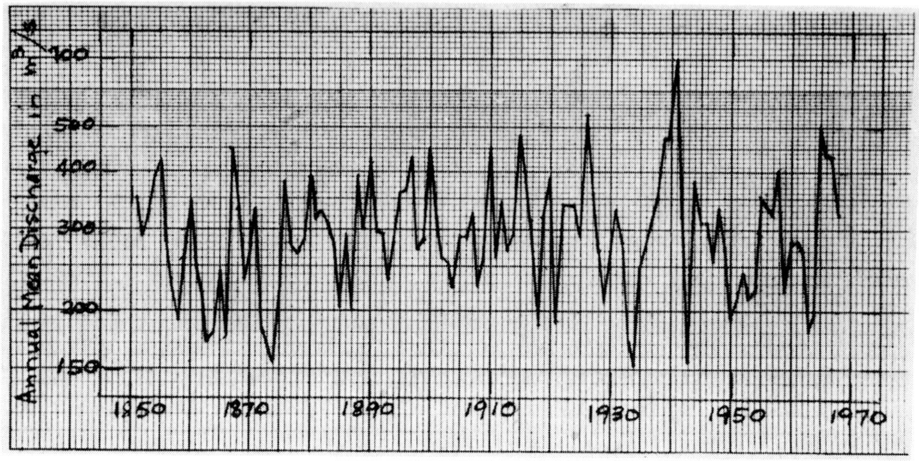

Fig. 10.2. Annual mean flows of Labe River, Czechoslovakia, 1851-1968.

This chapter describes effects on water resources of natural events and of man's activities, and concludes with some analytical methods of defining change. A broader geographic coverage of effects of man on water resources is given by Pereira (1973).

10.2 CHANGES DUE TO NATURAL EVENTS

Some observed changes in stream channels within the last 100 years or so might be interpreted as indications of climatic change. One such is the development, since about 1850, of many arroyos in southwestern United States. Opinions vary as to whether this was caused by precipitation changes or by livestock grazing, or both (Cooke and Reeves, 1976; Schumm and Hadley, 1957).

Schumm (1977) reported the destruction of the highly sinuous, relatively narrow and deep Cimarron River channel in southwestern Kansas by the major flood of 1914. Within the next 20 or 25 years of generally below average precipitation the channel widened from 50 to 1200 feet. Burkham (1972) documented the history of a portion of the channel of Gila River, Arizona. The channel was fairly stable and narrow from 1846 to 1904 and meandered through a flood plain. Average width was less than 150 feet in 1875 and less than 300 feet in 1904. During 1905-17 large floods which carried low sediment loads destroyed the flood plain and increased the channel width to about 2000 ft. The flood plain was reformed 1918-70 due to occurrence of only relatively low flood peaks carrying large sediment loads, and by the development of a dense cover of saltcedar. The average channel width was 200 feet in 1964 and 400 feet in 1968. The average annual flows for the periods 1918-64 and 1846-1904 were estimated to be about the same. Some other major changes in channel geometry resulting from major floods are described by Stevens, Simons, and Richardson (1975).

These channel changes may have been responses to changes in flow characteristics but homogeneous streamflow records are too short to verify this. An on-going program of gaging 57 natural streams indefinitely to provide the long records needed to assess natural changes is described in a report on Hydrologic Bench Marks by Cobb & Biesecker (1971).

Some natural events produce substantial changes in the hydrology of limited geographical regions. The earthquakes of 1811 caused land subsidence over several hundred square miles in Missouri and Tennessee, resulting in the formation of the 20-square mile Reelfoot Lake and many smaller lakes. Landslides caused by earthquakes may alter the topography; a major earthquake in western Montana in 1959 caused a landslide which formed a dam over 180 feet high on Madison River; water was backed up 6 miles. In addition to the disruption of flow below the landslide, prompt increases in streamflow were noted at a number of gaging stations on other streams in the area (Stermitz, 1964). Effects of the earthquake on streamflow were not detected after a few months, except on Madison River.

Hutchinson (1957, p. 42-47) describes lakes formed by landslides throughout the world. Some have existed for many years. Others were temporary and produced catastrophic floods when the dam failed. Landslides contribute much sediment to stream channels in some regions. Figure 10.3 shows a heavy accumulation of sediment resulting from a landslide on a small tributary to the Sun Kosi in Nepal.

Occasionally glacier movement causes storage and subsequent rapid release of water. The annual flood peaks of Knik River, Alaska, were greatly increased for at least 14 years because Knik Glacier advanced across Knik River each winter and temporarily stored the spring runoff. Destruction of the ice dam each summer by melting and erosion released the stored water causing a flood peak about an order of magnitude greater than would have occurred without the glacier blockage. See Figure 2.15 (Chapter 2). One of the largest floods of record occurred in 1922 on Iceland. The peak discharge of almost 2 million cubic feet per second resulted from a glacier outburst which released about 1.7 cubic miles of water over a 4-day period (Meier, 1969). Glacier-dammed lakes in other parts of the world are reported by Hutchinson (1957, p. 50-53).

Glaciers advance and retreat in response to small but persistent changes in climate. The runoff of a stream fed by a glacier will be reduced if the glacier is advancing and thus storing water as ice. Conversely a retreating glacier is releasing water and the runoff from it may exceed precipitation. Estimates of long-term flow characteristics of streams fed by glaciers should take into account the effect of glacier movement during the period of record.

Fig. 10.3. Effect of a landslide on the channel of a small tributary of Sun Kosi, Nepal.

Flow characteristics of a small stream can be substantially altered by construction of beaver dams. The impounded water promotes increased evaporation and ground-water recharge, albeit only temporarily; a flood may wash out the series of dams and return the stream to its previous condition except for reduced riparian timber.

Brush and forest fires often result from lightning strikes. The reduced plant cover and the modified soil cover resulting from a fire change the disposition of rain or snow on a basin. Interception is reduced but the change in infiltration depends on the original soil cover and on the soil characteristics. The rate and amount of runoff from a storm also depends on the basin topography.

On the San Dimas Experimental Forest in the San Gabriel Mountains of Southern California a 2.4 square mile watershed had about one-third of its vegetation destroyed by a wildfire in December 1953. Analysis of the precipitation and streamflow records for 17 years previous to and 2 years after the fire showed total storm discharges were three to five times as great after the fire as before, and that peak flows were increased by an order of magnitude or more because of the fire (Sinclair and Hamilton, 1955).

The effect of fire on the small steep basin in the San Dimas Experimental Forest is extreme. Storm runoff from larger basins with lower topographic relief and less intense rainfalls might show little or no effect from fire.

The natural event that produced the greatest hydrologic change in recent years was the volcanic eruption of Mt. St. Helens in Washington State (Findley, 1981). The volcano ejected large quantities of ash which melted the heavy snow pack. The resulting floods caused major destruction along the channels and carried the ash and other debris for many miles, Figure 10.4, finally clogging the shipping channel of Columbia River. After the flood receded the stream channels were left with deposits of ash (Fig. 10.5) which greatly reduced their carrying capacity. Thus what was previously considered a minor flood discharge on one of these channels will now inundate large areas along the stream and will move more material downstream and further reduce the channel capacity in the Toutle and Cowlitz Rivers.

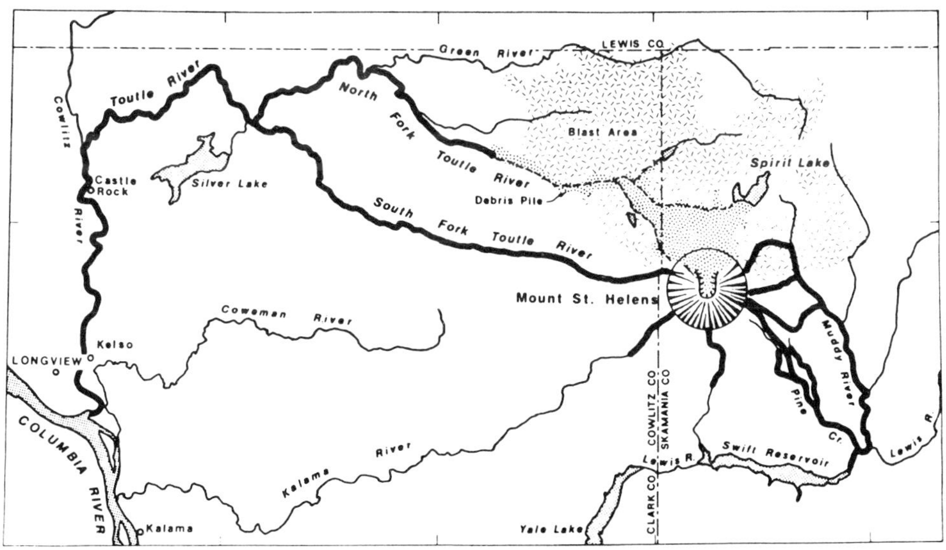

Fig. 10.4. Map of Mt. St. Helens area. Streams affected by major mudflows are indicated by heavy lines (From Meier and others, 1981).

In 1877 a major eruption of Mt. Cotopaxi, Ecuador, melted the snow and ice on its slopes and produced a flash flood and mud flow that inundated valleys as far as 200 miles away (Morrison, 1975).

Fig. 10.5. Channel filled with debris as a result of Mt. St. Helens eruption (USGS photo).

10.3 EFFECTS OF MAN'S ACTIVITIES

10.3.1 Surface storage

Dam construction increases evaporation losses, may increase ground-water recharge, and will change the flow regime downstream whether flow past the dam is regulated or not. The effect of increased evaporation due to reservoirs may be negligible in humid regions because that evaporation is balanced by the increased supply resulting from rainfall on the water surface rather than on the previous land surface. But in arid regions reservoir evaporation may limit the yield available from storage; beyond some reservoir capacity the loss from the increased reservoir surface area will be greater than the additional yield (Langbein, 1959).

Loss from a reservoir to ground water may be larger than that in the same reach before the reservoir existed. Ordinarily such losses are not large but as the residual in a water-budget equation, are difficult to estimate reliably.

If the ground-water table is in continuity with a stream whose level is raised by construction of a dam, then the water table will be raised accordingly. This may direct some ground-water flow to adjacent channels and may increase evapotranspiration from the ground water, both of which constitute a loss from storage.

The flow regimen below a storage reservoir depends on the purpose of the impoundment, the design, and the operation. The purpose of most impoundments is to change the flow regimen to provide flows when needed, or to reduce flood flows. Examples of changes due to impoundments are given in Figures 10.6, 10.7 and 10.8. Diurnal fluctuations, not shown in the examples, may be extreme if the flow is regulated for power generation.

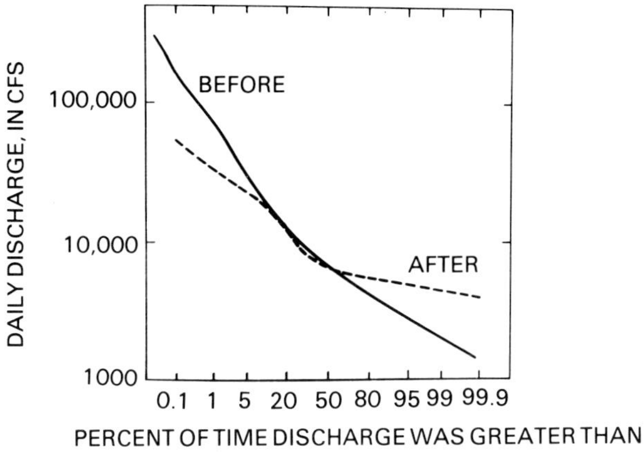

Fig. 10.6. Duration curves for Savannah River at Augusta, Georgia, before and after construction of Clark Hill Reservoir (From Stallings, 1967).

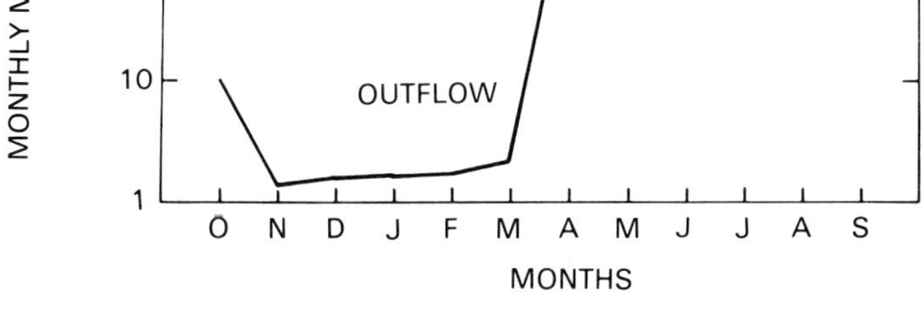

Fig. 10.7. Inflow and outflow of Owyhee Reservoir, Oregon, 1978 water year. Changes in reservoir contents and diversions for irrigation account for the differences.

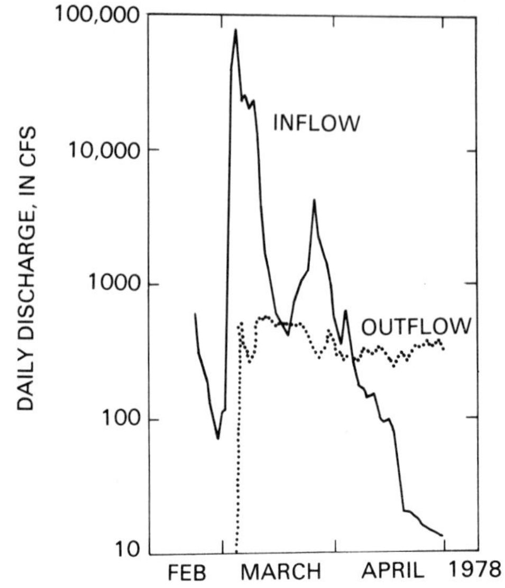

Fig. 10.8. Flow reduction due to Painted Rock flood-control reservoir, Gila River, Arizona.

Figure 10.9 shows the moderating effect of an SCS flood-water retarding structure under normal conditions. The effect will be slight if the flood occurs when the reservoir is nearly full. Reservoirs of this type are usually built on many small streams in a basin so that their aggregate effect is to substantially reduce the peak flows downstream. A series of reservoirs of this type was shown by Gilbert and Sauer (1970) to reduce the annual runoff appreciably for years of low inflow but only moderately for years of high inflow (Fig. 10.10).

Farm ponds and stock tanks (reservoirs) are built on many small streams but few of these are shown on maps. A study in the 1950s of the Cheyenne River basin (in Wyoming, South Dakota, and Nebraska) above Angostura Dam concluded that there were about 9,320 stock reservoirs with a storage capacity of 52,360 acre-feet, and that these reservoirs controlled an aggregate of 4,440 square miles. Based on 4-year records of inflow and outflow for about 50 reservoirs, the estimates of losses chargeable to all the reservoirs in the basin ranged from 19,000 acre-feet in a dry year to 80,000 acre-feet in a very wet year. Annual runoff from the basin ranges from 50,000 to 180,000 acre-feet (Culler, 1961).

Farm ponds in humid regions may not reduce streamflow yield appreciably because pond evaporation is compensated for by rainfall on the water surface. They may reduce flood peaks of small recurrence intervals.

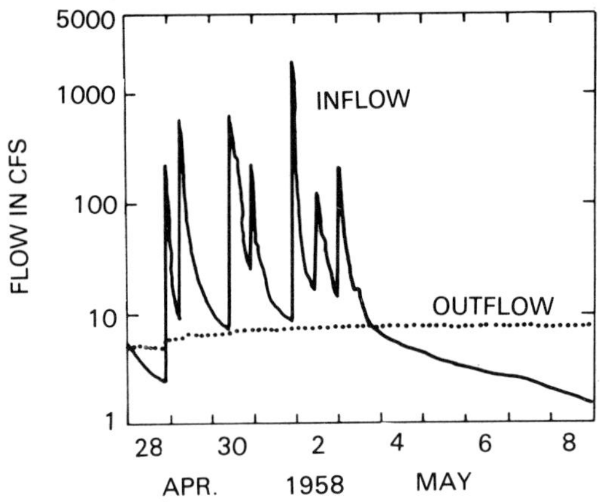

Fig. 10.9. Modification of flow by a fixed-outlet floodwater-retarding structure (From Gilbert and others, 1964).

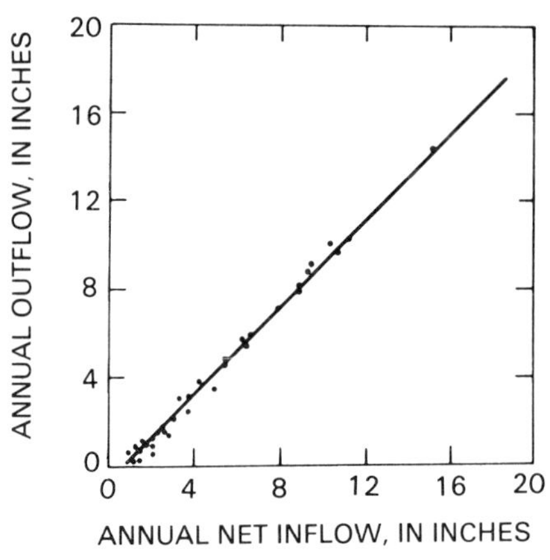

Fig. 10.10. Graph showing reduction in annual flow due to detention reservoirs in 7 study areas in Texas. Each study area had from 1 to 12 detention reservoirs (From Gilbert and Sauer, 1970).

Observations of runoff and sediment yield were made over a period of years at about 200 small reservoirs in the Missouri River, Colorado River, and Rio Grande basins from Montana to southern Arizona. The reservoirs ranged in capacity from 0.2 to 1,012 acre-feet and the contributing drainage areas ranged from 0.1 to 55 square miles. The reservoirs were selected to sample a wide range in climate, topography, geology, soil, and vegetation types. Consequently a wide variation in rates of runoff and sediment yield was found (Peterson, 1962). Spill from the reservoirs ranged from zero to as much as 95 percent of the inflow depending on the year, the size and character of the basin, and the size of the reservoir relative to the inflow.

Major changes in flow regimes result from flow diversions which may or may not be in conjunction with storage projects. Major diversions are usually measured or the flow below the diversion is gaged. Occasional measurements of a small diversion can be used to adjust the upstream monthly mean flows for the diversion.

An infrequent consequence of surface storage is a disastrous downstream flood due to dam failure or some other accident related to the storage project. The worst disaster of this type was the Johnstown, Pa. dam failure of 1889. The resulting wall of water 30 to 40 feet high cleaned out the downstream valley and took 2100 lives (Hoyt and Langbein, 1955). Some years ago an enormous wedge of earth slid into a reservoir in northern Italy raising the water level to well above the top of the dam and causing major damage and loss of life downstream. The more recent failure of Teton Dam in Idaho in 1976, and of some smaller structures in other States led to programs for evaluating the safety of existing dams in the United States. A similar program is being carried out in Great Britain.

The changed flow regime below a dam may affect the geometry of the channel and its biologic and esthetic character. Reservoir releases usually carry less sediment than the natural flow; consequently for some distance below the dam they pick up sediment by eroding the channel. The smaller flood peaks and possibly higher minimum flows usually result in adjustment to a smaller channel further downstream because of sediment deposits and encroachment by vegetation. Gregory and Park (1974) given an example. Schumm (1977) describes river metamorphoses resulting from flow regulation, and Williams (1978) documented changes in the channels of the North Platte and Platte Rivers in Nebraska, some of which are presently only 0.1 or 0.2 as wide as they were 100 years ago.

The quality of water released from deep reservoirs in temperate latitudes differs from that of the inflow not only in amount of suspended material but also in temperature and sometimes in dissolved solids concentration. The usual condition of thermal stratification in a reservoir during summer is shown in Figure 10.11. The temperature of released water will depend on the elevation of the intake. Since construction of Glen Canyon Dam on Colorado River, Arizona, water temperatures at Lees Ferry, a few miles downstream, are higher in winter and cooler in summer than before.

If chemical stratification occurs in a reservoir the quality of the water released also depends on the depth of the intake. The distribution of dissolved solids in Flaming Gorge Reservoir, Wyoming-Utah, shows concentrations ranging from 450 mg/l at depths of 100 feet to 700 mg/l near the 400-foot depth (Madison and Waddell, 1973). The mean concentration of dissolved solids in Flaming Gorge Reservoir is significantly greater than that of the inflows because of leaching from the inundated areas.

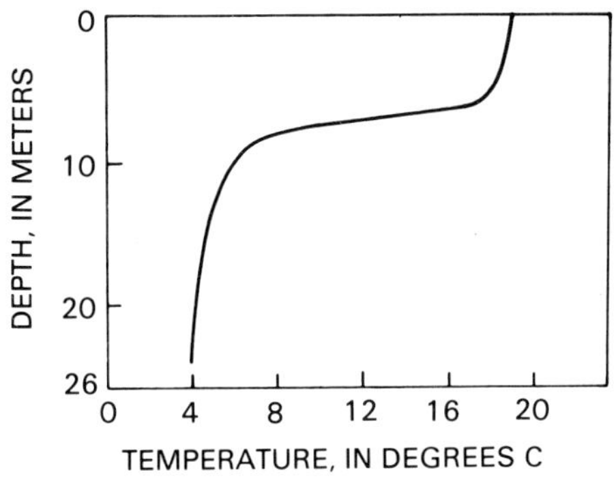

Figure 10.11. Typical vertical distribution of lake temperature during summer.

Prior to 1964 when the Aswan High Dam went into operation, floods on the Nile River removed domestic, industrial, and agricultural wastes annually. Since then floods no longer occur; the flow follows requirements for irrigation and for hydropower generation from the Dam. Consequently the River below the Dam is accumulating pollutants in spite of a 1962 law prohibiting disposal of industrial wastes in public water bodies unless certain requirements are met. (Mobarek, 1980).

10.3.2 Land-use changes

Modification of the land surface and of the vegetal cover affects the hydrology of a basin in varying degrees depending on the magnitude and areal extent of the change and on the physical characteristics of the basin and its climate. The difficulty of identifying changes in the hydrology attributable to small or moderate changes in the basin accounts for some of the conflicting reports in the literature. Conclusions from one valid experiment may not apply to other regions or basins. Some general statements about the qualitative effects of certain land treatments, however, are widely applicable and are useful in planning and design of water-resources projects.

(i) Timber lands. The management of timber lands for both wood and water requires an understanding of how logging and reforestation affect streamflow. Changes may occur in annual yield, summer low flow, magnitude and regimen of floods, and distribution of flow throughout the year. In general, removal of trees reduces interception and evapotranspiration and permits more of the precipitation to reach the streams. In humid northeastern United States removal of all vegetation from a forested watershed will increase annual water yield the first year after treatment by 4 to 12 inches (Lull and Reinhart, 1967). The wide range of these increases is attributed to the methods of logging and to climatic differences among basins. In such humid regions regrowth of trees and shrubs begins soon after logging. Consequently the streamflow yield trends back toward the precutting values although substantial increases may be maintained for years. This was demonstrated by Swank and Helvey (1970) who recorded the annual water yield from a small basin that was clearcut in 1939 and again in 1962. During the second regrowth period (after 1962) streamflow increases after the first year were about half the increases at the same points in time after the first treatment (Swift and Swank, 1981).

Clearcutting part of a basin should result in an increase proportional to the percentage of area logged; selective cutting probably would have a lesser effect. Cutting only 10 to 20 percent of a basin usually produces effects too small to be detected.

Flood flows may be increased by logging but the change is usually difficult to detect. Schneider and Ayer (1961) analyzed runoff from two small central New York basins, one agricultural and the other being reforested. They concluded that the flood peak magnitudes decreased on the reforested basin as the trees grew and that spring floods occurred later. See section 10.4.1 for their method of analysis. Analysis of the same data by Satterlund and Eschner (1965) revealed that runoff during snowmelt periods is gradually concentrated in a shorter time following reforestation of open lands and that there is a gradual desynchronization with runoff from the agricultural watershed so that the concentrated snowmelt runoff comes later in the season in the years following reforestation.

Hewlett and Helvey (1970) reported that clearcutting a mature hardwood forest in the southern Appalachians caused increases of about 11 percent in stormflow volumes. Peak discharges increased slightly after logging but the statistical significance test for peaks was not as conclusive as that for volumes.

Selective timber removal has been found to affect snow accumulation, the timing of snowmelt, the water yield, and the peak runoff timing and magnitude but the effects vary widely on various topographic sites (Anderson and others, 1958).

Anderson (1960) described various research management studies and concluded that there is no single way of managing land in the snow zone to improve all aspects of water yield: more water, better timing of water yield, better quality of water, and reduced floods and sediment movement.

Replacement of hardwoods with pine in eastern United States resulted, after 10 years, in a reduction in water yield, mostly in the dormant season, probably because of greater interception loss. Results of experiments in tropical Africa on water use by tall rain forest, tea plantations, bamboo, and pine and cypress are given by Pereira (1973).

Base flows can be substantially increased by removing trees and shrubs adjacent to the stream channels but cutting on hilltops and at intermediate elevations well above the water table has little effect. Dunford and Fletcher (1947) reported that cutting stream-bank vegetation on a 22-acre watershed in the Coweeta Experimental Forest in North Carolina virtually eliminated the diurnal fluctuation during the growing season and increased daily mean flows from 4 to 19 percent in a recession period. A study of low flows in the Rappahannock River basin in Virginia led to the following conclusions: (1) that the differences among low flows (per unit area) of the nine streams studied are related to differences in the proportions of cleared land in the basins, (2) that the cleared land, mostly in pasture, generally produces higher summer and fall base flows than timbered land, (3) that the land-clearing effects are proportionally greatest at the lower discharge rates, and (4) that for the purpose of increasing summer and fall base flows, clearing is most effective adjacent to stream channels (Riggs, 1965). Many other studies, some of which are reported by Johnson and Meginnis (1960), substantiate these general conclusions.

(ii) Rangelands. Grass is the important resource in rangelands. Water yield from such lands in the United States is generally low because of little precipitation or because the topography is not favorable for producing runoff. On other continents, notably Africa, considerable runoff and ground-water recharge may occur seasonally.

Grazing by livestock and game animals reduces the grass cover and increases soil exposure both of which increase the runoff. Grazing also reduces infiltration rates (Gifford & Hawkins, 1978). Pereira (1973, p. 142-146) describes the effects of grazing animals on watersheds in various climatic and topographic regimes.

An intensive study of the effect of grazing on the hydrologic and biotic characteristics of small drainage basins on the Badger Wash basin in the Colorado Plateau of western United States was begun in 1953. From 1954-66 runoff from the grazed watersheds averaged about 33 acre-feet per square mile per year; runoff from ungrazed ones averaged about 70 percent of that from the grazed watersheds. Sediment yield from grazed watersheds was about 3 acre-feet per square mile per year and about 2 for the ungrazed watersheds (Lusby, Reid, and Knipe, 1971). The terrain in which this experiment was conducted is shown in Figure 10.12.

Fig. 10.12. General view of the terrain of the Badger Wash basin, Colorado (From Lusby and others, 1971).

The effect of range fire is the same as that of severe overgrazing in semiarid regions. In watersheds with deeply-rooted grasses, the old grass may be burned off annually to promote growth of new grass and to prevent growth of brush. Here fire usually has little effect on water or sediment yield except temporarily.

Sagebrush is being replaced by crested-wheat grass in some more-or-less level rangelands in western United States where annual precipitation is as little as 10 inches. Runoff from such lands is negligible whether in native or replacement vegetation.

Trees such as mesquite and juniper are removed from semiarid regions of southwestern United States in order to promote the growth of grass for livestock. Any change in water yield is masked by the high variabiliy of the ephemeral flows.

(iii) Croplands. Considerable information is available on the effects of various cultivation and cropping practices on runoff from plots of a few acres. Results of a comprehensive investigation in Nebraska was reported by Sharp, Gibbs, and Owen (1966). They found that conservation practices reduced surface runoff from small watersheds from 25 to 40 percent, particularly in dry years, but that no satisfactory method was available for transferring such results to large complex watersheds. They tried numerous methods of evaluating the effects of watershed treatment on streamflow but none would consistently assess such affects on large basins, or even prove conclusively that such effects do or do not exist.

On small plots the effects of changing from one cropping practice to another are variable and sometimes opposite depending on the particular combination of soils, geology, climate, vegetation, and water management practices.

(iv) Urbanization. Construction of buildings, streets, parking lots, and storm drainage systems increases the amount of runoff by reducing infiltration, and the rate of runoff because of increased velocity overland and in the improved channels. Changes in the runoff hydrograph resulting from complete urbanization of a basin are shown symbolically in Figure 10.13. The increase in peak flow depends on

1. The recurrence interval of the flood. Those of large recurrence intervals will be increased less than those of low recurrence intervals because heavy, protracted rainfalls temporarily reduce the infiltration of a natural basin.
2. The degree of development, and the percent of the basin developed.
3. The infiltration characteristics of the natural basin. Increased runoff from impervious surfaces depends on the change in infiltration. Paving over a sandy soil will produce a much greater increase in runoff than paving over a tight clay.
4. The design of the drainage system.
5. Whether the development includes provision for temporary storage on roofs, in yards and parking lots, and in reservoirs. Provision of storage is required in some jurisdictions. Ordinances in the Chicago area require that the stormwater runoff rate after development be no more than it was before development; Talhami (1980) describes methods used to meet this requirement by temporary on-site detention.

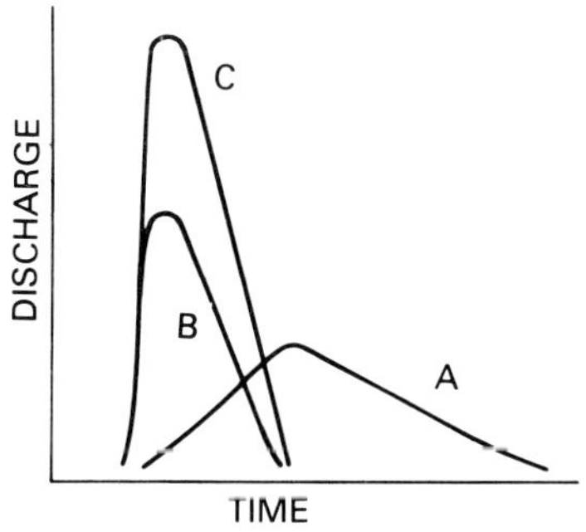

Fig. 10.13. Flood hydrographs showing the effect of urbanization. A, rural; B, basin sewered; C, basin sewered and with an impervious surface (After Anderson, 1970).

Storm runoff data have been collected and analyzed at only a few urbanized areas in the United States. Some results are given by Anderson (1970) for

Northern Virginia; Martens (1968) for metropolitan Charlotte, N.C.; Dempster (1974) for Dallas, Texas, metropolitan area; Liscum and Massey (1980) for Houston, Texas, metropolitan area; and Wibben (1977) for Nashville, Tennessee. These results show ratios of flood-peak discharges from urbanized basins to those from undeveloped basins ranging from about 1 for Nashville residential development to about 5 in Houston and Northern Virginia. The low ratios are for areas having soils with low infiltration rates. Drainage-system improvement, or its lack, also accounts for some of the variation.

Leopold (1968) used available information to develop a generalized relation for the effects of various degrees of urbanization on the mean annual flood (Fig. 10.14). That graph gives the ratio by which the mean annual flood peak from an undeveloped basin will be increased because of urbanization. The peak for the undeveloped condition can be estimated from a regression on basin characteristics (Chapter 8).

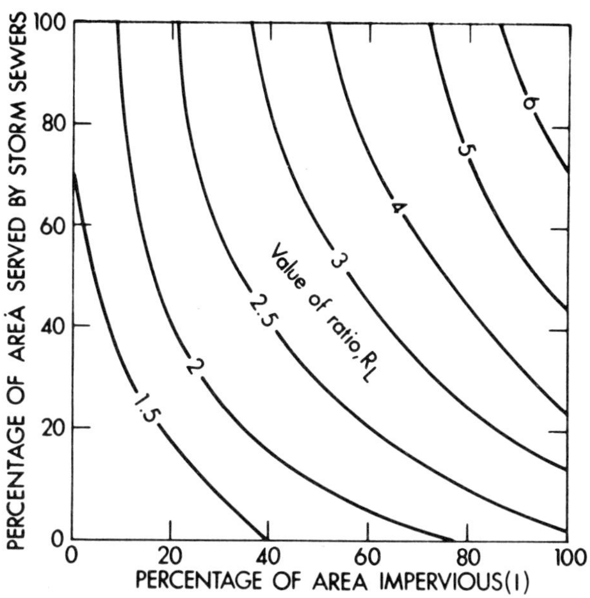

Fig. 10.14. Graph showing ratio of urban to rural mean annual floods (From Leopold, 1968).

Espey and Winslow (1974) used data from East Coast, Texas, Mississippi, Michigan, and Illinois watersheds to derive regression equations for urban floods of various recurrence intervals. One of their equations is

$$Q_{50} = 297\ A^{0.85}\ I^{0.22}\ S^{0.50}\ R^{1.57}\ \phi^{-1.61}$$

where Q_{50} is 50-year flood in cubic feet per second, A is drainage area in square miles, I is impervious cover as a percentage, S is dimensionless channel slope, R is 6-hour rainfall at 50-year recurrence interval, and ϕ is a factor depending on channel conditions. A, S, and R are estimators of the peak flood from the undeveloped basin; I, and ϕ account for the effect of urbanization.

Sauer and others (1983) compiled a nationwide urban flood data base and developed equations for estimating urban flood flow characteristics in ungaged areas. Their equations have standard errors ranging from 37 to 52 percent and are applicable throughout the conterminous United States and Hawaii.

Urbanization may affect ground-water recharge and low flows but whether an increase, decrease, or no change, occurs after urbanization depends on local conditions. Urbanization generally has little effect on the low flows of streams because the area made impervious rarely is a significant percentage of the drainage area of a basin large enough to produce an appreciable low flow under natural conditions.

Runoff from urban areas contains dissolved and suspended waste materials that may degrade the water quality of the receiving streams. The possibility of having water-quality problems as a consequence of urbanization depends on climate, the type of recipient water body, and other factors. In many places urban runoff is not likely to result in water-quality problems (EPA, 1982).

(v) Irrigation. Irrigation is a means of providing water to plants during periods of inadequate rainfall. Water may be pumped from the ground or diverted from a stream. Pumping of ground water affects streamflow only when it removes water that would otherwise have gone to the stream. If a well is close to a stream, the effect of ground water pumping may be the same as a stream diversion. At some distance from the stream the immediate reduction will be less than the rate pumped. Figure 10.15 shows the hydrograph of stream depletion resulting from pumping 10 acre-feet per day for 35 days from an alluvial aquifer at a point 3600 feet from the Arkansas River in Colorado. Ground water pumping does not affect streamflow in regions where the water table is always below the streambeds. Other effects of ground-water pumping are described in a following section.

More water than the plants need is often applied to the land when surface water is used for irrigation. The excess either seeps into the ground and raises the water table or is drained by surface channels to a stream. The principal effects of irrigation on streamflow are upstream regulation and diversions and the impaired quality of the water draining from the project.

The salinity of the drainage water is usually greater than that of the applied water because of leaching of salts from the soil. In flat areas artificial drainage is usually provided; if not, the water table may rise nearly to the land surface and the consequent evaporation of ground water will concentrate the salts near the soil surface, severely reducing plant growth.

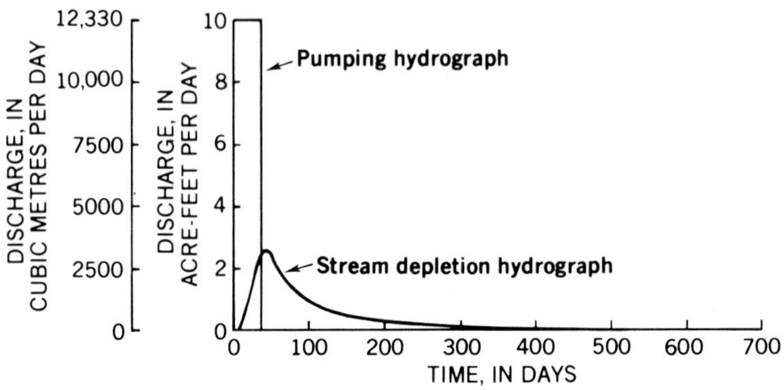

Fig. 10.15. Example of residual effects of well pumping for 35 days (From Jenkins, 1968).

A succession of irrigation projects along a river progressively reduces the flow and increases the salinity. Notable examples of streams on which this has occurred are Colorado River (Iorns, Hembree, and Oakland, 1965) and River Murray, Australia (Hart, 1980).

Irrigation diversions from streams feeding terminal lakes can reduce those lakes in volume and surface area. The level of the Caspian Sea has been reduced in recent years because of irrigation and other consumptive uses of water (Faizi, 1980). Likewise the Dead Sea level has been declining because of the reduced inflows of Jordan River and other tributaries (Sauer, 1978). Terminal lakes or swamps (sinks) in western United States can be greatly changed, or eliminated, by reduction of natural inflows.

Water spreading is another form of irrigation. Streamflow is diverted by low dams onto the flood plains or the riparian land. In the mountainous western United States the spring runoff is used to irrigate hay meadows. In semiarid parts of the world the channels of small ephemeral streams are terraced so that the flows, except during major floods, are all absorbed in the terraced area (Fig. 10.16).

Introduction of water to dry soil may produce land subsidence through hydrocompaction. Some of the subsidence in the San Joaquin Valley, California is attributed to hydrocompaction. (See the section on Ground Water Pumping).

(vi) Drainage and Channelization. Proper drainage is essential for many land uses besides agriculture. The hydrological effects of draining swamp lands depend on the geologic and topographic settings. The outflow from a wetland is

Fig. 10.16. Terraced land in the channel of a wadi in North Yemen.

dominated by the slow movement of water through the wetland, and by the high evapotranspiration rate. Drainage of a wetland results in faster removal of water following precipitation, and greatly reduced evapotranspiration because of the reduction in water surface area. These changes may result in increases in low flows and in annual runoffs. Klueva (1975), in a study of 16 basins in Byelorussia ranging in area from 67 to 8770 km^2, reported that the minimum daily winter mean discharge in 8 basins increased after reclamation (drainage by channelization) by an average of 20 to 130 percent. In the other basins under study no significant changes were found. Reclamation in all basins under study resulted in an increase of streamflow in the summer and autumn period.

Similar changes were found by Mustonen and Seuna (1975) who calibrated 2 natural basins for 22 years; calibration was continued for 9 years after 40 percent of one basin was drained for forestry. Drainage produced an increase of 43 percent in annual runoff and increases of many times in winter and summer annual minimums. Heath (1975) showed that deepening and widening of the natural channels in the swampy 57-square mile drainage basin of Ahoskie Creek, North Carolina increased the minimum flow from zero to about 2 cfs.

Although the results quoted above are consistent, different effects may be found in other types of basins. See Pereira (1973, p. 194-200).

(vii) Ground-water pumping. Where ground water drains to streams and helps maintain their flows, any extraction of ground water will reduce the low flows and the yields. An example of the effect of temporary pumping was shown in Figure 10.15. The magnitude of the effect on streamflow characteristics of ground-water extraction depends on the hydraulic connection between the aquifer and the stream and on the pumping schedule and the amount pumped.

Substantial lowering of the water table or the piezometric surface in certain types of soils results in land subsidence. Some parts of the San Joaquin Valley, California have subsided 9 meters due to ground-water withdrawal. Substantial subsidence has also occurred in the Houston-Galveston (Texas) area, and in Japan, Venice (Italy) and various other places in the world. The maximum known subsidence due to ground-water pumping occurred in Mexico City, parts of which have been lowered as much as 18.7 meters in 80 years. Rebounds of a few centimeters have been reported from Japan and Venice after recovery of artesian pressures. See IAHS (1976) and Poland (1981).

Land subsidence may change stream gradients, induce more recharge or more evaporation from the channel, and increase the riparian areas subject to flooding in addition to its adverse effects on buildings, highways, canals, and on agriculture.

(viii) Phreatophyte removal. Phreatophytes are deep-rooted plants that draw their moisture primarily from ground water; common ones are saltcedar, arrowweed, cottonwood, saltbush, willow, and mesquite. In the southwestern United States these plants extract large quantities of water from the shallow ground water along streams and adjacent valleys. The most common and one of the highest users of water is saltcedar (tamarix), an introduced plant, which did not appear on the flood plains until the 1920s. By 1965 the plant infested about a million acres of western stream valleys.

The possibility of reclaiming water used by such nonbeneficial vegetation led to studies to determine the amount of water used, and how some of it might be salvaged. A study by Gatewood and others (1950) concluded that phreatophytes used about 28,000 acre-feet of water in a 12-month period in a 46-mile reach of Gila River in Arizona. A six-year observation of water used by saltcedar growing in evaporation tanks has shown how water use is related to plant density and depth to water table (Van Hylckama, 1970). Computations of potential water salvage are reported by Hughes and McDonald (1966).

A 9-year study of a 15-mile reach of the Gila River, Arizona, was begun in 1962 (1) to evaluate water salvage by phreatophyte control on a flood plain typical of areas of existing and proposed application, (2) to describe the hydrological and ecological variables of the study area sufficiently for extrapolation of water salvage data to other sites, and (3) to develop methods for evaluating hydrologic variables in a large area. The water-budget method was used to estimate evapotranspiration from 5500 acres of the flood plain, then the flood plain was cleared of vegetation and the evapotranspiration again estimated. Results show that the average yearly consumptive use of 43 inches before clearing was reduced to 19 inches after clearing (Culler and others, 1982).

Water formerly used by riparian vegetation can be salvaged only if regrowth is prevented or the water table is lowered by pumping, neither of which may be practical.

10.3.3 Cloud seeding

Cloud seeding as a means of inducing or augmenting precipitation or of modifying its character was proposed in the late 1940s. Schaefer (1946) showed that precipitation could be induced by providing nuclei for ice formation in supercooled clouds. He used dry ice as the source of the nuclei. Silver iodide particles, proposed by Vonnegut (1947), as a seeding agent, are now generally used. They are introduced into the atmosphere by "generators" located on mountain ridges where the wind will carry the particles into the clouds. Guidelines for cloud seeding to augment precipitation are described by ASCE (1983).

The principal objective of most cloud seeding is to provide rain for crops. Although it has been clearly demonstrated that seeding a suitable cloud will induce precipitation, the effectiveness of operational cloud seeding is controversial. Warner (1974) said that seeding clouds with a view of stimulating additional rainfall over areas of a few thousand square kilometers has been done for a quarter of a century and many programs in operation are considered successful, but when we come to examine the scientific basis for this assumption, there is no general agreement as to whether, or in what conditions, it is justified. Neyman and Scott (1974) claim that experimental designs based on area-to-area comparisons are unreliable and that the questions raise problems of mathematical statistics that are fundamentally new. WMO recently summarized the conclusions of a panel of experts by, "Most of the past weather modification experiments are considered inconclusive by the scientific community. Careful evaluations involving both extensive physical measurements of the clouds and precipitation and statistical analyses of the measurements are now considered mandatory in arriving at sound judgments. At this time weather modification other than supercooled fog dispersal must be considered in the realm of research. Enhancing precipitation or suppressing hail reliably and on demand remain distant goals."

Cloud seeding is a controversial activity beyond the question of its effectiveness. Induction or augmentation of precipitation at a particular time may benefit some crops and damage others; within a region there may be no agreement on the desirability of weather modification at a particular time. Even attempts to increase the winter snowpack in the Colorado Rockies by cloud seeding were met with some objections.

10.3.4 Air pollution

Inadvertent cloud seeding by particles introduced to the atmosphere from industrial activities may produce local increases in precipitation. Huff (1977) reported on a 5-year field and analyses program at St. Louis, Missouri, where 225 recording rain gages were operated on 5200 km^2. Data indicated approximately a 30-percent increase in summer rainfall in an area that is most frequently exposed to storms crossing the urban-industrial region. The enhancement was most pronounced in rainstorms that produced maxima of 25 mm or more in the network. The average water yield for individual raincells in those storms was over 70 percent greater in urban-exposed cells than in surrounding rural cells. The study indicates that the frequency distributions of heavy rainfalls of 5-minute to 2-hour duration may vary significantly between urban, suburban, and rural areas in large urban-industrial regions. This variation probably should be considered in defining flood characteristics from small watersheds subject to rainfall enhancement although adequate data are rarely available.

Some particles introduced to the atmosphere have effects beyond that of augmenting precipitation. Man-made oxides from smelters, fossil-fueled power plants, other industrial activities, and automobile engine exhausts react with moisture to produce a precipitation containing corrosive acids. Or the pollutants may fall to earth in dry form and be transformed into corrosive acids by subsequent precipitation or by melting of snow. This acid rain increases the acidity of lakes and streams, except those in regions rich in limestone. Many lakes in northeastern United States, Canada, and Scandinavia no longer contain fish because acid rain has lowered the pH of the water below 5.

Acid rain can affect plants, animals, and inanimate objects. Prior to regulation, smelter effluents, mainly sulfur dioxide, were responsible for killing timber on large downwind areas. In a few years before and after 1900, effluent from the smelters at Copper Hill, Tennessee, completely denuded 17,000 acres of forest land and injured 30,000 additional acres. The smelter at Trail, B.C., Canada destroyed timber for 40 miles down the Columbia River valley in the 1920s. A Landsat view of the Wawa, Ontario, Canada region clearly shows the 350 square mile area on which vegetation was killed by sulfur dioxide emissions from a sintering plant (NASA, 1976, p. 199). Recent effects of acid rain are less severe but much more widespread (LaBastille, 1981).

10.4 QUANTIFYING EFFECTS OF CHANGES

10.4.1 Environmental changes

An experiment could be conducted on a single basin on which precipitation and surface runoff are measured for a number of years until an acceptable calibration is defined between the flow characteristic of interest and the precipitation input. Then the change that is to be evaluated in the basin would be made and data collection continued until a range of precipitation inputs have been recorded. The effect of the basin change is the difference between the measured output after change and the output for that period computed from the pretreatment calibration.

The sensitivity of this procedure depends on the size of the standard error of the calibration. The effect of the change must appreciably exceed the standard error if it is to be statistically significant. A smaller real effect may exist although it cannot be shown by this technique.

Disadvantages of this experimental basin approach are the cost of years of data collection, the long delay between beginning of the experiment and the answer sought, and the lack of a firm basis for applying the result to other than the study basin. Slivitzky and Hendler (1965) describe the difficulties of using experimental basins and conclude that in most cases the limitations outweigh the advantages. Hewlett, Lull, and Reinhart (1969) discuss the pros and cons of the experimental basin approach and justify its use for some objectives.

A variation of the experimental basin is the paired-basin approach in which two similar, nearby basins are gaged until a reasonable range in discharge is obtained. The discharge of the one that is to be modified is regressed on the discharge of the other. Then changes are made in the first basin and data collection on both is continued for a few years. The Wagon Wheel Gap experiment, designed and carried out by Bates and Henry (1928) is an example; Hoyt and Troxell (1934) analyzed data from this experiment. The paired contiguous watersheds of about 200 acres each, located in the Rio Grande drainage basin of Colorado at altitudes of between 9000 and 11000 feet, were geologically and topographically similar and subject to the same meteorological conditions because of their small size. Forest cover also was similar. Reliable observations of runoff and precipitation were made for 8 years in the natural state and for 7 years after complete deforestation of one basin. Results are shown in Figure 10.17. All annual runoffs of basin A for years subsequent to deforestation are significantly greater statistically, at the 95% level, than before. One might fit a regression line to the after-deforestation points but this would require an assumption that the condition of basin A remained unchanged after deforestation; this assumption is unlikely because of regrowth, and in fact the plotted points seem to indicate a change with time. Van Haveren (1981) used data from the Wagon Wheel Gap experiment to show the effect of deforestation on annual, peak, and low flows and on hydrograph timing. Another example of an analysis using paired basins is given by Harris (1977).

The effect of a gradual basin change such as that due to reforestation can be identified by relating annual flows from the basin being reforested to flows from a control basin and to number of years from beginning of experiment. Schneider and Ayer (1961) used the regression model

$$R_o = a + b_1 R_a - b_2\ T\ R_a$$

where R_s is runoff from the reforested basin, R_a is runoff from the control basin, T is position of the year in the numerical sequence beginning with T = 1 for the first year of record. The form of the equation indicates that the magnitude of the change in R_s is dependent on the magnitude of R_a as well as on the time sequence. For the period 1939–57 reductions in runoff from the small reforested basin in central New York averaged about 0.36 inch per year.

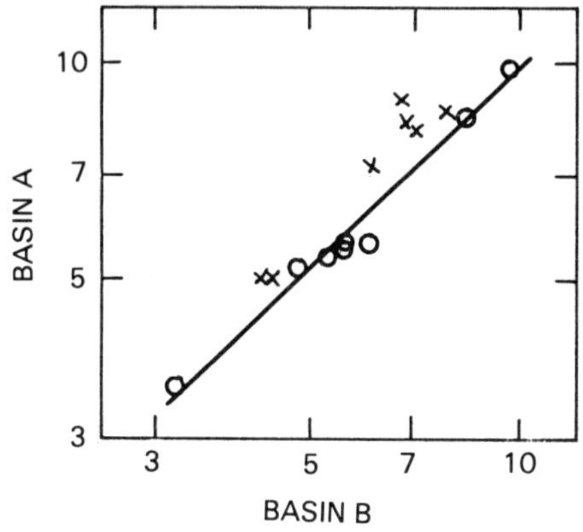

Fig. 10.17. Paired basin method of defining effect of deforestation. Graph shows annual runoffs in inches. Circles are for calibration period; crosses are results after deforestation of Basin A.

Changes that have occurred may also be quantified by use of a mathematical model but the principal use of such models is for estimating changes that will result from some proposed modification or development in a basin. A model is a quantitative representation of the system being studied. After calibration with basin data, the model may be used to predict the output from a given input, or to compute the change in output due to some modification of the basin. Both single-event and continuous watershed models are used to detect change. Ordinarily the change indicated by modeling is not verified by data; confidence in the result is based on how well (1) the model structure conforms to the system, (2) the parameters describe specific elements of the system, particularly those elements that will be changed, and (3) the calibration data fit the model. The fundamentals of stormwater modeling, and use of models for evaluating the effects of land use on flood flows are given by Overton and Meadows (1976). Feldman (1980) evaluated the capability of some commonly-used hydrologic models for predicting peak discharges of prescribed frequencies under various basin conditions. Feldman (1981) described theory and experience with HEC models for water resources system simulation.

An approach used to define the effects of urbanization on flood peaks consists of gaging a dozen or more small streams whose basins encompass a range from rural to fully urbanized. The flood frequency characteristics are defined from the gaged records, sometimes supplemented by information derived by a rainfall-runoff model. Regression of a flood frequency characteristic on basin characteristics, including an index of urbanization, is developed. Then the changes due to urbanization can be determined by solving the equation for various values of the index of urbanization. If the region covered by the gaged basins was reasonably homogeneous hydrologically before urbanization, the changes defined by this method should apply throughout that region. For more details see the section on Urbanization in this chapter.

One of the simpler SCS (1975) models uses soils data, channel length and character, and storm rainfall of a specific recurrence interval to compute, for the urbanized condition, the peak discharge having the same recurrence interval as the rainfall.

More complex models are needed for urban hydrology where lumped parameters do not adequately describe the physical system, where many alternative development plans are to be evaluated, and when various types of answers are sought. A guide to urban model analysis is given by ASCE (1977).

10.4.2 Diversion and regulation

Major changes in streamflow characteristics result from storage and diversion of flow. The modified flow characteristics can be defined by a combination of hydrologic and hydraulic models given suitable information on the pattern of operation of the reservoirs. Jeffcoat and others (1974, 1976) developed a stream-reservoir model to simulate the operation of the Lakes Marion-Moultrie reservoir system in South Carolina. The model incorporates an inflow component, a diversion component, and a reservoir system component and is capable of evaluating the performance of the reservoir system for a variety of operating rules. The calibrated model was used to simulate 31 years of outflow from the inflow record under an assumed operating rule. Flow characteristics were defined from the simulated record.

REFERENCES

Anderson, D.G., 1970, Effects of urban development on floods in northern Virginia: U.S. Geol Survey Water-Supply Paper 2001-C, 22 p.

Anderson, H.W., Rice, R.M., and West, A.J., 1958, Snow in forest openings and forest stands: Soc. American Foresters Proc., p. 46-50.

Anderson, H.W., 1960, Prospects for affecting the quantity and timing of water yield through snowpack management in California: Western Snow Conf. Proc., 28th Annual Meeting, Santa Fe, N.M., April 12-14, 1960, p. 44-50.

ASCE, 1977, Guide for collection, analysis, and use of urban stormwater data: Am. Soc. of Civil Engineers Conference Rept., 115 p.

ASCE, 1983, Guidelines for cloud seeding to augment precipitation, Jour. of Irrigation and Drainage Div., ASCE, Vol. 109, No. IRI, p. 111-182.

Bates, C.G. and Henry, A.J., 1928, Forest and streamflow experiment at Wagon Wheel Gap, Colorado: U.S. Weather Bureau Monthly Weather Review, Supplement 30, 79 p.

Burkham, D.E., 1972, Channel changes of the Gila River in Safford Valley, Arizona, 1846-1970: U.S. Geol. Survey Prof. Paper 655-G, 24 p.

Cobb, E.D. and Biesecker, J.E., 1971, The national hydrologic-benchmark network: U.S. Geol. Survey Circ. 460-D, 38 p.

Cooke, R.U. and Reeves, R.W., 1976, Arroyos and environmental change in the American Southwest: Oxford, England, Clarendon Press.

Culler, R.C., 1961, Hydrology of stock-water reservoirs in upper Cheyenne River basin: U.S. Geol. Survey Water-Supply Paper 1531-A, 136 p.

Culler, R.C., Hanson, R.L., Myrick, R.M., Turner, R.M., and Kipple, F.P., 1982, Evapotranspiration before and after clearing phreatophytes, Gila River flood plain, Graham County, Arizona: U.S. Geol. Survey Prof. Paper 655-P, 67 p.

Dempster, G.R., 1974, Effects of urbanization on floods in the Dallas, Texas, metropolitan area: U.S. Geol. Survey Water Resources Investigations 60-73.

Dunford, E.G. and Fletcher, P.W., 1947, Effect of removal of streambank vegetation upon water yield: Am. Geophys. Union Trans., Vol. 28, p. 105-110.

EPA, 1982, Preliminary results of the nationwide urban runoff program, Executive Summary: Water Planning Division, U.S. Environmental Protection Agency, Washington, D.C.

Espey, W.H., Jr. and Winslow, D.E., 1974, Urban flood-frequency characteristics: Jour. of Hydraulics Division, ASCE, Vol. 100, No. HY2, p. 279-293.

Faizi, S., 1980, Hydrographic changes in the area of the Caspian Sea, USSR, in Pollution and Water Resources, Columbia University Seminar Series, Vol. XIII, Part 2, p. 109-129: Elmsford, New York, Pergamon Press.

Feldman, A.D., 1980, Flood hydrograph and peak-flow frequency analysis, in Improved Hydrologic Forecasting, Why and How: Am. Society of Civil Engineers, p. 1-19.

Feldman, A.D., 1981, HEC models for water resources system simulation, theory and experience in Chow, V.T., Advances in Hydroscience, Vol. 12, New York, Academic Press, p. 297-423.

Findley, R., 1981, Mount St. Helens: National Geographic, Vol. 159, No. 1, January 1981.

Gatewood, J.S., Robinson, T.W., Colby, B.R., Hem, J.D., and Halpenny, L.C., 1950, Use of water by bottomland vegetation in lower Safford Valley, Arizona: U.S. Geol. Survey Water-Supply Paper 1103, 210 p.

Gifford, G.F. and Hawkins, R.H., 1978, Hydrologic impact of grazing on infiltration: A critical review: Water Resources Research, Vol. 14, No. 2, p. 305-313.

Gilbert, C.R., Commons, G.G., Koberg, G.E., and Kennon, F.W., 1964, Hydrologic studies of small watersheds, Honey Creek basin, Collin and Grayson Counties, Texas, 1953-59: U.S. Geol. Survey Water-Supply Paper 1779-F.

Gilbert C.R. and Sauer, S.P., 1970, Hydrologic effects of flood-water retarding structures on Garza-Little Elm Reservoir, Texas: U.S. Geol. Survey Water-Supply Paper 1984, 95 p.

Gregory, K.J. and Park, C., 1974, Adjustment of river channel capacity downstream from a reservoir: Water Resources Research, Vol. 10, No. 4, p. 870-873.

Harris, D.D., 1977, Hydrologic changes after logging in two small Oregon coastal watersheds: U.S. Geol. Survey Water-Supply Paper 2037, 31 p.

Hart, B.T., 1980, Water-quality management in arid and semiarid areas: CETESB Seminar on Water Resources of Arid and Semiarid Regions, Recife, Brasil, March 16-22, 1980.

Heath, R.C., 1975, Hydrology of the Albemarle-Pamlico region, North Carolina: U.S. Geol. Survey Water Resources Investigations 9-75, 48 p.

Hewlett, J.D., Lull, H.W., and Reinhart, K.G., 1969, In defense of experimental watersheds: Water Resources Research, Vol. 5, No. 1, p. 306-316.

Hewlett, J.D. and Helvey, J.D., 1970, Effects of forest clear-felling on the storm hydrograph: Water Resources Research, Vol. 6, No. 3, p. 768-782.

Hoyt, W.G. and Troxell, H.C., 1934, Forests and streamflow: Trans. Am. Soc. of Civil Engineers, Vol. 99, p. 1-30.

Hoyt, W.G. and Langbein, W.B., 1955, Floods: Princeton, N.J., Princeton University Press, 469 p.

Huff, F.A., 1977, Urban effects of storm rainfall in midwestern United States: International Assoc. of Hydrological Sciences Publ. No. 123, Proc. of the Amsterdam Symposium, p. 12-19.

Hughes, G.H., and McDonald, C.C., 1966, Determination of water use by phreatophytes and hydrophytes: Jour. of Hydraulics Division, ASCE, Vol. 92, No. HY2, p. 63-81.

Hutchinson, G.E., 1957, A treatise on limnology, Vol. 1: New York, John Wiley and Sons, 1015 p.

IAHS, 1976, Land Subsidence: International Assoc. of Hydrological Sciences Publ. No. 121, Proc. of the Anaheim Symposium, 669 p.

Iorns, W.V., Hembree, C.H., and Oakland, G.L., 1965, Water resources of the upper Colorado River basin - Technical Report: U.S. Geol. Survey Prof. Paper 441-A, 370 p.

Jeffcoat, H.H., Jennings, M.E., and Peterson, J.B., 1974, Lakes Marion-Moultrie stream-system investigation: Part I - model selection, calibration, and error analysis: U.S. Geol. Survey Water-Resources Investigations 25-74, 55 p.

Jeffcoat, H.H., Jennings, M.E., Collins, D.L., and Shearman, J.O., 1976, Lakes Marion-Moultrie stream-system investigation, Part II - simulation studies: U.S. Geol. Survey Water-Resources Investigations 76-11, 25 p.

Jenkins, C.T., 1968, Techniques for computing rate and volume of stream depletion by wells: Ground Water, Vol. 6, No. 2, p. 37-46.

Johnson, E.A. and Meginnis, H.G., 1960, Effect of altering forest vegetation on low flows of small streams: International Assoc. of Scientific Hydrology Publ. No. 51, General Assembly of Helsinki, Commission of Surface Waters, p. 257-266.

Klueva, K.A., 1975, The effect of land reclamation by drainage on the regime of rivers in Byelorussia: Hydrology of Marsh-Ridden Areas, Proc. of the Minsk Symposium, June 1972, The UNESCO Press, Paris, p. 419-436.

LaBastille, A., 1981, Acid rain, how great a menace?: National Geographic, Vol. 160, No. 5, November 1981, p. 652-681.

Langbein, W.B., 1959, Water yield and reservoir storage in the United States: U.S. Geol. Survey Circ. 409.

Leopold, L.B., 1968, Hydrology for urban land planning - A guidebook on the hydrologic effects of urban land use: U.S. Geol Survey Circ. 554, 18 p.

Liscum, F. and Massey, B.C., 1980, Technique for estimating the magnitude and frequency of floods in the Houston, Texas metropolitan area: U.S. Geol. Survey Water-Resources Investigations 80-17, 29 p.

Lull, H.W. and Reinhart, K.G., 1967, Increasing water yield in the northeast by management of forested watersheds: U.S. Forest Service Research Paper NE-66, 45 p.

Lusby, G.C., Reid, V.H., and Knipe, O.D.,. 1971, Effects of grazing on the hydrology and biology of the Badger Wash basin in western Colorado: U.S. Geol. Survey Water-Supply Paper 1532-D.

Madison, R.J. and Waddell, K.M., 1973, Chemical quality of surface water in the Flaming Gorge Reservoir area, Wyoming and Utah: U.S. Geol Survey Water-Supply Paper 2009-C, 18 p.

Martens, L.A., 1968, Flood inundation and effects of urbanization in metropolitan Charlotte, North Carolina: U.S. Geol. Survey Water-Supply Paper 1591-C.

Meier, M.F., 1969, Glaciers and water supply: Jour. American Water Works Assoc., Vol. 61, No. 1, January 1969, p. 8-12.

Meier, M.F., Carpenter, P.J., and Janda, R.J., 1981, Hydrologic effects of Mt. St. Helens' 1980 eruptions: EOS, Trans. Am. Geophysical Union, Vol. 62, No. 23, Aug 18, 1981.

Mobarek, I.E., 1980, Nile hydrology and water quality: CETESB Seminar on Water Resources of Arid and Semiarid Regions, Recife, Brasil, March 17-21, 1980.

Morrison, T., 1975, The Andes; The World's Wild Places: Alexandria, Va. Time-Life Books.

Mustonen, S.E. and Seuna, P., 1975, Influence of forest drainage on the hydrology of an open bog in Finland: Hydrology of Marsh-Ridden Areas, Proc. of the Minsk Symposium, June 1972, The UNESCO Press, Paris, p. 519-529.

NASA, 1976, Mission to earth: LANDSAT views the world: National Aeronautics and Space Administration, Washington, D.C.

Neyman, J. and Scott, E.L., 1974, Rain simulation experiments: design and evaluation, in Proc. of WMO/IAMAP Scientific Conference on Weather Modification, World Meteorological Organization Publ. No. 399, Geneva, 538 p.

Overton, D.E. and Meadows, M.E., 1976, Stormwater modeling: New York, Academic Press, 358 p.

Pereira, H.C., 1973, Land use and water resources: London, Cambridge University Press, 246 p.

Peterson, H.V., 1962, Hydrology of small watersheds in western States: U.S. Geol. Survey Water-Supply Paper 1475-I.

Poland, J.F., 1981, Subsidence in United States due to ground-water withdrawal: Jour. of Irrigation and Drainage Division, ASCE, Vol. 107, No. IR2, June 1981, p. 115-135.

Quinn, F.H., 1981, Secular changes in annual and seasonal Great Lakes precipitation, 1854-1979, and their implications for Great Lakes water resource studies: Water Resources Research, Vol. 17, No. 6, p. 1619-1624, December 1981.

Riggs, H.C., 1965, Effect of land use on the low flow of streams in Rappahannock County, Virginia: U.S. Geol. Survey Prof. Paper 525-C, p. 196-198.

Satterlund, D.R. and Eschner, A.R., 1965, Land use, snow, and streamflow regimen in central New York: Water Resources Research, Vol. 1, No. 3, p. 397-405.

Sauer, S.P., 1978, Recent and projected changes in Dead Sea level and effects on mineral production from the Sea: U.S. Geol. Survey Open-File Rept. 78-176, 43 p.

Sauer, V.B., Thomas, W.O., Jr., Stricker, V.A., and Wilson, K.V., 1983, Flood characteristics of urban watersheds in the United States: U.S. Geol. Survey Water-Supply Paper 2207, 63 p.

Schaefer, V.J., 1946, The production of ice crystals in a cloud of supercooled water droplets: Science, 104, 457.

Schneider, W.J. and Ayer, G.R., 1961, Effect of reforestation on streamflow in central New York: U.S. Geol. Survey Water-Supply Paper 1602, 61 p.

Schumm, S.A., 1977, The fluvial system: New York, John Wiley & Sons, (p. 159-164).

Schumm, S.A. and Hadley, R.F., 1957, Arroyos and the semiarid cycle of erosion: Am. Jour. of Science, Vol. 255, March 1957, p. 161-174.

SCS, 1975, Urban hydrology for small watersheds: U.S. Soil Conservation Service Tech Release No. 55, Washington, D.C.

Sharp, A.L., Gibbs, A.E., and Owen, W.J., 1966, Development of a procedure for estimating the effects of land and watershed treatment on streamflow: U.S. Dept. of Agriculture Tech. Bull. No. 1352, 57 p.

Sinclair, J.D. and Hamilton, E.L., 1955, Streamflow reactions of a fire-damaged watershed: Proc. ASCE, Hydraulics Division, Vol. 81, Separate No. 629, 17 p.

Slivitzky, M.S. and Hendler, M., 1965, Watershed research as a basis for water resources development: National Research Council of Canada, Proc. of Hydrology Symposium No. 4, Research Watersheds, p. 289-294.

Stallings, J.S., 1967, South Carolina streamflow characteristics, low-flow frequency and flow duration: U.S. Geol. Survey Open-File Rept., 83 p.

Stermitz, F., 1964, Effects of the Hebgen Lake earthquake on surface water: U.S. Geol. Survey Prof. Paper 435-L, p. 139-150.

Stevens, M.A., Simons, D.B., and Richardson, E.V., 1975, Nonequilibrium river form: Jour. of Hydraulics Division, ASCE, Vol. 101, No. HY5, p. 557-566.

Swank, W.T. and Helvey, J.D., 1970, Reduction in streamflow increases following regrowth of clearcut hardwood forests: International Association of Hydrological Sciences Publ. No. 96, p. 346-360.

Swift, L.W., Jr. and Swank, W.T., 1981, Long term responses of streamflow following clearcutting and regrowth: Hydrological Sciences Bull., Vol. 26, No. 3, September 1981, p. 245-256.

Talhami, A., 1980, Temporary detention cuts storm flow peaks: Civil Engineering, Vol. 50, No. 12, December 1980, p. 72-75.

Van Haveren, B.P., 1981, Wagon Wheel Gap watershed experiment revisited: Western Snow Conf. Proc., Apr. 14-16, 1981, p. 131-138.

Van Hylckama, T.E.A., 1970, Water use by saltcedar: Water Resources Resarch, Vol. 6, No. 3, p. 728-735.

Vonnegut, B., 1947, The nucleation of ice formation by silver iodide: Jour. Applied Physics, 18, 593.

Warner, J., 1974, Rain enhancement: A review, in Proc. of WMO/IAMAP Scientific Conf. on Weather Modification, Tashkent, 1-7 October 1973: World Meteorological Organization (WMO-No. 399), Geneva, 538 p.

Wibben, H.C., 1977, Effects of urbanization on flood characteristics in Nashville-Davidson County, Tennessee: U.S. Geol. Survey Water-Resources Investigations 76-121.

Williams, G.P., 1978, The case of the shrinking channels - the North Platte and Platte Rivers in Nebraska: U.S. Geol. Survey Circ. 781, 48 p.

Chapter 11

APPLICATIONS OF HYDROLOGIC DATA

11.1 INTRODUCTION

Planning, design, and operation of projects that involve or affect water resources require reliable information. This chapter describes some of the common engineering problems and the techniques available for their solution using hydrologic information developed in previous chapters. The techniques given here are relatively simple, utilize available data, and produce generally reliable results.

11.2 RESERVOIR DESIGN

Reservoirs are formed to moderate the range in streamflows. The objective of providing storage is most often to provide a dependable flow greater than the natural low flow. Another important objective is to reduce flood peaks, and a common one is to provide a lake for recreation or esthetic reasons. Only the hydrologic aspects of reservoir design are treated here, principally what storage is required to provide a target outflow for a given inflow. Some aspects of the hydrologic design such as evaporation losses and reservoir sedimentation, cannot be quantified until a specific location for the reservoir has been proposed and surveyed. Reservoir location and dam design are the province of engineers and geologists. In areas of low topographic relief, suitable reservoir sites may not be available.

11.2.1 Mass curve

Storage analyses for water supply have traditionally been based on the mass curve of streamflow. Consider a long streamflow record as the inflow to a reservoir. Cumulative daily, weekly, or monthly volumes from the inflow record are plotted against time to produce the mass curve. Figure 11.1 shows a short portion of a mass curve based on monthly values. Any desired constant draft rate can be represented by a straight line of appropriate slope. For example the line for a draft rate of 200 cfs would rise 12 x 200 = 2400 cfs months per year.

Interpretation of the mass curve is based on the following assumptions: (1) that the reservoir was full at the beginning of the streamflow record, and (2) that as long as there is water in the reservoir the demand will be met in full. In Figure 11.1 the most deficient period of flow begins in late 1941 and extends to the middle of 1943. The storage required to furnish 200 cfs throughout this period is represented by the maximum vertical distance between the mass curve and the 200 cfs draft rate line. Smaller draft rates could have been sustained during this period with smaller storages. The maximum realistic draft rate is always less than the mean flow for the period of record.

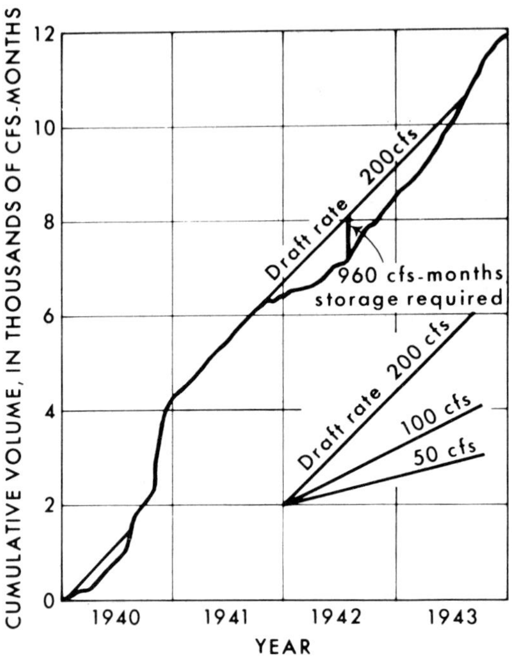

Fig. 11.1. Mass curve of monthly volumes, Blackwater River, Virginia.

Figure 11.1 shows only four years of record. Estimates of storage required to sustain a given draft rate usually increase in reliability with length of streamflow record used.

The mass curve can also be analyzed for a draft rate that varies throughout the year by preparing the appropriate use line.

11.2.2 Use of simulated streamflows

The mass-curve method does not permit assignment of a reliable probability of failure to the derived storage. A single record, say one of 25 years, may define the mean flow of the stream reasonably well but it provides only one sequence from which to compute storage required. Other sequences of monthly volumes having the same means and standard deviations would produce different storages. Such sequences can be derived by simulation. Typically 100 sets of 50-year sequences might be generated. Each set would be analyzed by the mass-curve method or its equivalent to determine 100 values of the maximum storage required in 50 years for each of several draft rates. Assuming a design life of

50 years for the reservoir, the probability of failure to provide the design draft rate can be related to the volume of storage by a frequency analysis of the storage volumes.

The use of synthetic records in storage analysis was reported by Thomas and Fiering (1962). Burges and Linsley (1971) evaluated the method. It has been widely used for important projects where streamflow records are available.

11.2.3 Annual mass-curve method

The mass-curve method can be used to define the storage required for a given draft rate for each year of record. From these storages a set of draft-storage-frequency curves can be prepared. This approach is limited to draft rates that can be sustained by the streamflow available in any one year; that is, by within-year storage. The usefulness of this analysis depends on the annual variability of streamflow. In some regions, the maximum draft that can be provided by within-year storage is less than a tenth of the mean flow. In others, notably the southeastern United States, drafts of half or more of the mean flow can be provided by within-year storage.

Daily, rather than monthly, discharges should be cumulated. In many parts of the United States the year beginning April 1 is appropriate because a reservoir would most likely be full on that date.

The computation and plotting of a mass curve on a daily basis is time consuming. The process of cumulation can be simplified by an arithmetic analysis using monthly values except during the critical dates of high and low storage. Table 11.1 is an example of such an analysis. Note that it is necessary to check whether the reservoir would refill by March 31. In the example (see footnote 2) it did not. Thus the draft rate of 100 cfs would require over-year storage and the analysis of within-year storage should be limited to a smaller draft rate.

Even analyses of the type shown in Table 11.1 are time consuming if several draft rates are studied but the analyses can be made quickly by computer.

Table 11.2 shows output of a computer program for 4 draft rates and for the same gaging-station record analyzed in Table 11.1. Underlined figures indicate that the tabulated storage required was not replenished by the following April 1 and therefore the corresponding draft cannot be sustained by within-year storage alone.

Frequency curves of storage required to maintain draft rates of 20, 40, and 60 cfs are shown in Figure 11.2. Only the higher half or third of the storages computed in Table 11.2 need be used because the lower parts of the frequency curves are not of interest. No curve for a draft of 100 cfs is shown because that draft could not have been provided from within-year storage in 1943 and 1949. A draft rate of 80 cfs should have been included.

TABLE 11.1

Mass analysis of daily discharges for Moosup River at Moosup, Conn., April 1, 1943, to March 31, 1944 [Draft rate is 100 cfs]

Month	Total flow in cfs-days	Draft in cfs-days	Surplus in cfs-days	Cumulative Surplus plus 10,000 in in cfs-days
Apr	5,821	3,000	2,821	12,821
May	6,998	3,100	3,898	16,719
June 1-4	577	400	177	16,896
5-30	1,521	2,600	-1,079	15,817
July	1,072	3,100	-2,028	13,789
Aug	772	3,100	-2,328	11,461
Sept	375	3,000	-2,625	8,836
Oct	1,263	3,100	-1,837	6,999
Nov	2,952	3,000	-48	6,951
Dec	1,734	3,100	-1,366	5,585
Jan	1,861	3,100	-1,239	4,346
Feb.1-14	641	1,400	-759	3,587
15-29	2,407	1,500	+907	4,494
Mar	7,594	3,100	4,494	8,988

[1]Storage required: 16,896 (high flow) - 3,587 (low flow) = 13,309 cfs-days.
[2]Deficiency on March 31: 13,309 (storage required) - [8,988 (Mar. 31) - 3,587 (low flow)] = 7,908 cfs-days.

11.2.4 Probability routing

An alternative to generating a long simulated streamflow record for defining the required storage is provided by a combination of the probability-routing method and the annual-mass-curve method. Results of the probability-routing method, after adjustment for seasonal variations, define that part of the draft-storage-frequency relation which depends on over-year storage. The annual-mass-curve method is used to define the relation for lower draft rates.

Probability routing is applied to a distribution of annual discharges (inflows) under the assumptions that the discharge for each year is uniform throughout that year and is equal to the mean for that year and that the annual discharges are not serially related. The procedure was proposed by Langbein (1958) and was described by Hardison and Furness (1963).

Briefly stated, probability routing is based on assuming a distribution for start-of-year reservoir contents and then, for a given draft rate and given frequency distribution of inflow, computing the distribution of year-end reservoir contents to see if it checks the assumed start-of-the-year distribution. (The end-of-year probability distribution of reservoir contents is inherently equal to the start-of-year distribution.) The computations are begun by subdividing the reservoir capacity into about 15 layers and assuming a probability for each layer as well as a probability of spill and a probability of being empty.

TABLE 11.2

Output of computer program for within-year storage, Moosup River at Moosup, Conn. (underlined figures indicate that the tabulated storage was not replenished by the following April 1)

	Storage, in cfs-days, required to maintain the draft rates indicated during year beginning April 1			
Year	20 cfs	40 cfs	60 cfs	100 cfs
1933	36	568	1,928	5,207
1934	82	978	2,385	5,689
1935	170	1,614	3,979	11,409
1936	80	1,333	3,054	9,989
1937	20	315	1,251	5,997
1938	10	41	116	799
1939	25	761	2,195	8,076
1940	19	371	2,156	6,436
1941	361	1,939	5,023	12,813
1942	25	412	1,445	6,000
1943	319	1,912	4,438	13,308
1944	301	1,501	2,926	6,511
1945	62	1,557	3,854	9,108
1946	24	282	1,062	5,617
1947	85	926	2,897	8,239
1948	113	1,124	2,708	7,317
1949	139	2,157	5,157	12,469
1950	83	1,115	3,697	9,790
1951	34	652	2,660	7,756
1952	44	1,293	3,903	10,677
1953	378	2,153	4,580	11,009
1954	21	274	1,347	4,716
1955	29	370	1,010	2,735
1956	81	1,253	3,772	10,079
1957	1,092	4,127	8,005	16,535
1958	14	54	269	1,625
1959	145	1,143	2,731	6,328

The year-end probability for each state is computed by multiplying the assumed probability of each of the possible states by the deficiency probability of the inflow required to produce the selected year-end state and cumulating the products. Agreement could be reached by trial and error solution starting each time with the previously computed year-end distribution, but a more feasible way is to solve 15 to 20 simultaneous equations by computer to obtain the unique distribution of year-end reservoir contents for the given storage capacity and given draft rate. Each of these equations gives the probability of the water level being in one of 15 to 20 states at the end of the year and the solution is based on the fact that the sum of the probabilities must equal unity. The probability of the reservoir being empty at the end of the year thus obtained is the desired information for the draft rate, reservoir capacity, and inflow distribution used in setting up the equations.

Probability routing can be performed for any distribution of annual inflows, but by characterizing the inflow distributions by a few parameters, the results

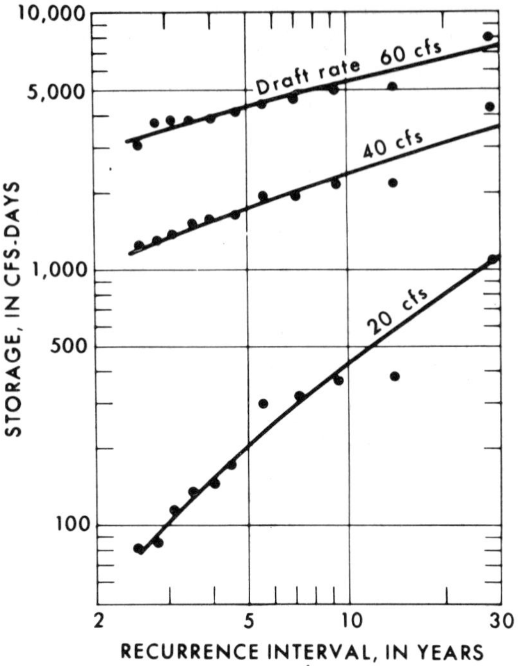

Fig. 11.2. Draft-storage-frequency curves for Moosup River, Connecticut (for within-year storage).

can be generalized so that it is not necessary to resort to the computer for each problem. Solutions have been obtained for three types of two-parameter distributions of inflow and the results have been related to a variability index. A sample of the generalized computer solutions is given in Table 11.3. These storage requirements are based on annual inflows and must be increased to account for the within-year flow variability. Riggs and Hardison (1973) give the complete tables of the computer solutions and describe how the distribution of annual inflows and the frequency curves of within-year storage (by the annual mass-curve method) are used to prepare a draft-storage relation such as that of Figure 11.3. In that figure, within-year storage will provide a dependable draft of only about 20 percent of the mean flow.

11.2.5 Evaporation, sedimentation, and bank-storage

Good estimates of evaporation losses from a reservoir cannot be made until the water-surface area is known. Evaporation losses may appreciably reduce the draft available from a given storage, especially in arid regions. An analysis by Leopold (1959) concludes that increases in storage in the Colorado River basin beyond a certain volume will achieve practically no additional water

TABLE 11.3

Carryover storage requirements for normal distributions of annual flows (Storage requirements are in ratio to mean annual runoff; C_v is coefficient of variation of annual flows)

C_v	Draft, in percent of mean flow										
	100	98	95	90	80	70	60	50	40	30	20
				5-percent chance of deficiency							
.08	1.00	.25	.11	.04							
.10	1.25	.36	.18	.08							
.12	1.50	.50	.26	.12	0						
.14	1.75	.65	.34	.17	.03						
.16	2.01	.81	.43	.23	.06						
.18	2.26	.97	.53	.29	.10						
.20	2.52	1.14	.64	.35	.14	.02					
.22	2.77	1.32	.76	.42	.18	.06					
.24	3.03	1.51	.88	.49	.23	.10	0				
.26	3.28	1.70	1.01	.57	.28	.14	.03				
.28	3.53	1.90	1.14	.65	.34	.18	.06				
.30	3.79	2.10	1.28	.74	.40	.22	.09	0			
.40		3.17	2.02	1.27	.71	.47	.30	.16	.06		
.50		4.30	2.83	1.89	1.07	.75	.53	.37	.24	.13	.03
				1-percent chance of deficiency							
.08		.48	.21	.10							
.10		.73	.30	.16	.03						
.12		1.02	.43	.24	.08						
.14		1.33	.60	.32	.14	.03					
.16		1.69	.80	.41	.19	.08					
.18		2.09	1.02	.51	.25	.13	.01				
.20		2.52	1.24	.62	.32	.19	.06				
.22		2.97	1.47	.76	.39	.24	.12	.01			
.24		3.45	1.71	.91	.46	.30	.17	.06			
.26			1.96	1.08	.54	.36	.23	.11	0		
.28			2.23	1.26	.62	.42	.28	.16	.05		
.30			2.50	1.45	.71	.48	.33	.21	.10	0	
.40			3.96	2.48	1.28	.87	.64	.48	.36	.25	.11
.50				3.65	2.04	1.32	1.01	.78	.65	.52	.32

regulation if evaporation loss is subtracted. This concept of a practical maximum net flow from storage was utilized by Lof and Hardison (1966) in computing storage requirements for water in the United States. Rates of evaporation from water surfaces are published; see Chapter 12.

The storage capacity of a reservoir will be progressively reduced by deposition of sediment carried by the stream. Estimates of the annual volume of deposited material can be made from records of sediment transport and estimates of the trap efficiency, that is, the proportion of sediments trapped by the reservoir; see Vanoni (1975, p. 587-602). In regions where sediment-transport data are not available, periodic surveys of existing reservoirs will provide useful information. The economic life of a reservoir depends on the rate of reduction in storage capacity.

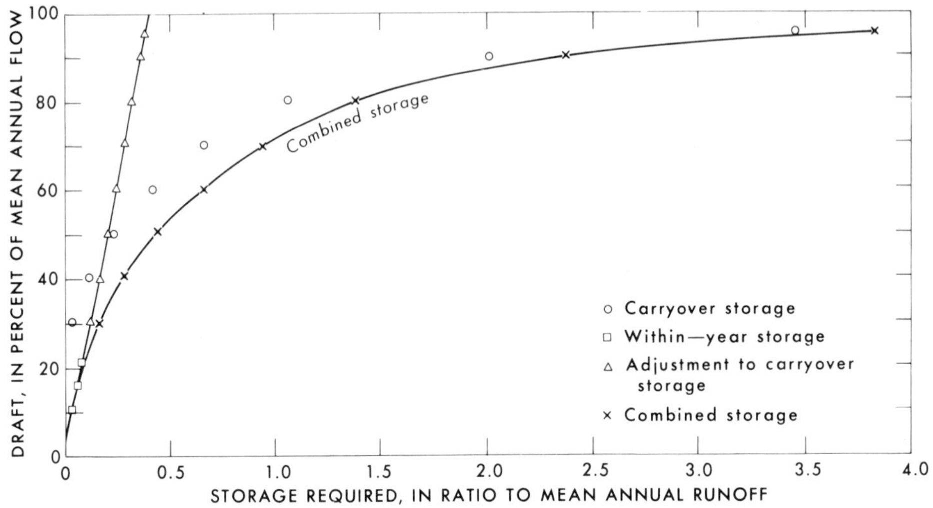

Fig. 11.3. Draft-storage curve for 2 percent probability of deficiency, Red River, Tennessee.

Reservoir storage capacity is sometimes found to be appreciably greater than that computed from the topographic survey. This increase is due to bank storage. Water flows into the banks as the reservoir stage rises and much of it returns on falling stages. Thus, except for a residual in the bank aquifers, this water is not lost. See Bank Storage in Chapter 6.

11.2.6 Draft-storage at ungaged sites

Definition of draft-storage-frequency relations at ungaged sites is often required. This might be done by estimating the statistics of monthly flows from generalized relations with basin characteristics, synthesizing a long record from these, and analyzing that record. It is much simpler to transfer draft-storage-frequency characteristics directly from gaged to ungaged sites.

Transferring a draft-storage relation is accomplished by use of a regional relation in conjunction with some streamflow information collected or estimated at the site. Two criteria must be met: (1) draft-storage relations defined by gaging-station records for various recurrence intervals must be closely related to flow characteristics; (2) sufficient information must be available at the site of application to provide good estimates of the flow characteristics needed in the regional relation.

Regional draft-storage relations are usually developed graphically because some of the models are difficult to express mathematically. The choice of variables will depend on flow characteristics in the region and on what information is available. However, the draft and the storage should be in units commonly used or readily computed.

Various relations are shown in Figures 11.4, 11.5 and 11.6. Drainage area is required in all of them. The other variables, a low-flow characteristic, mean flow, and an index of variability of annual mean flows, may be estimated at ungaged sites by methods given in previous chapters.

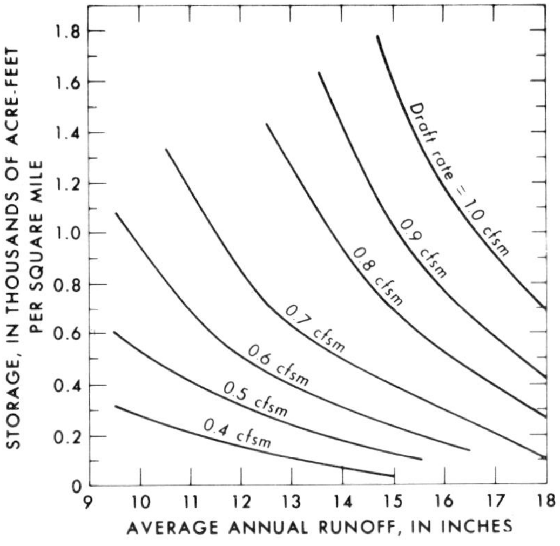

Fig. 11.4. Draft-storage curves for a region in Missouri; 2 percent chance of deficiency (From Skelton, 1971).

11.2.7 Spillway design floods

Spillways are designed to protect the dam and reservoir against failure. The degree of protection provided should depend on the consequences of failure. For high dams whose failure would cause loss of life and extensive property damage, the spillway is usually designed to pass the maximum reasonably possible flood. Where dam failure would cause minor damage downstream the spillway may be designed to pass a lesser flood.

Streamflow records are too short to provide an adequate basis for defining the probable maximum flood by frequency analysis, and none of the theoretical distributions used approach a limit. But meteorologists accept the premise of a probable maximum storm precipitation, from which a maximum probable flood can be computed. Procedures for estimating the probable maximum precipitation are

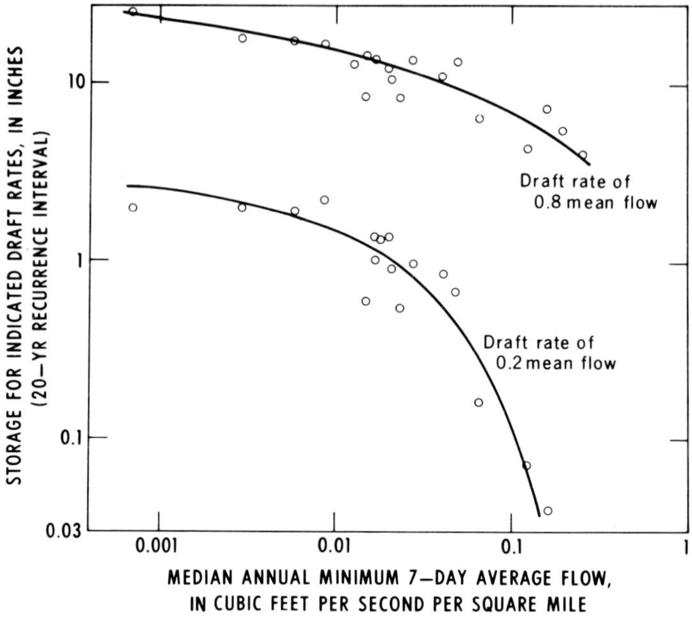

Fig. 11.5. Regional draft-storage relations based on data for some Illinois streams.

described by Hershfield (1961) and by WMO (1973). Methods of converting probable maximum precipitation into extreme flood flows are given by WMO (1969b). Probable maximum precipitation as a basis for determining spillway design floods was evaluated by an ASCE Task Force on the hydrology of spillway design (Banks, 1964). The Task Force (1) reviewed and summarized existing policy and criteria for spillway design floods for dams of all sizes, (2) examined hydrologic considerations, policies, criteria, and methodology for determination of design flood flows, (3) described policies, criteria and methodology for determination of spillway capacities in relation to design flood flows, and (4) evaluated the accuracy of statistical analyses of hydrologic data pertinent to flood flows.

11.2.8 Storage for flood control

The ideal flood-control reservoir is empty most of the time so that its capacity can be used to reduce the incoming flood peaks. An approximate method of defining the storage required to limit outflow to some target discharge is shown in Figure 11.7. The hydrograph is commonly derived from a project storm but it may be that of a major recorded flood. Release need not be at the maximum rate after the inflow falls below that rate but the storage is usually reduced promptly.

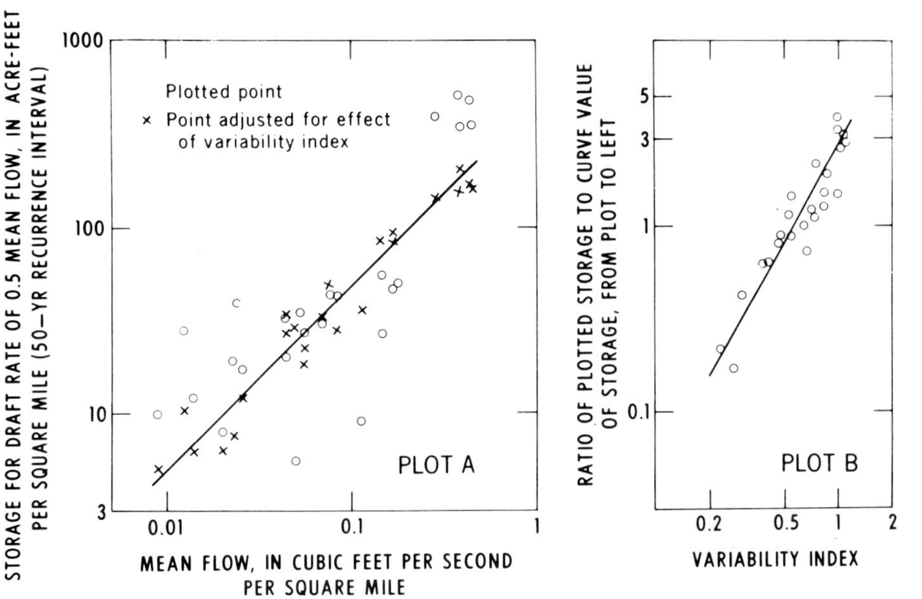

Fig. 11.6. Storage related to mean flow and a variability index in Kansas River basin.

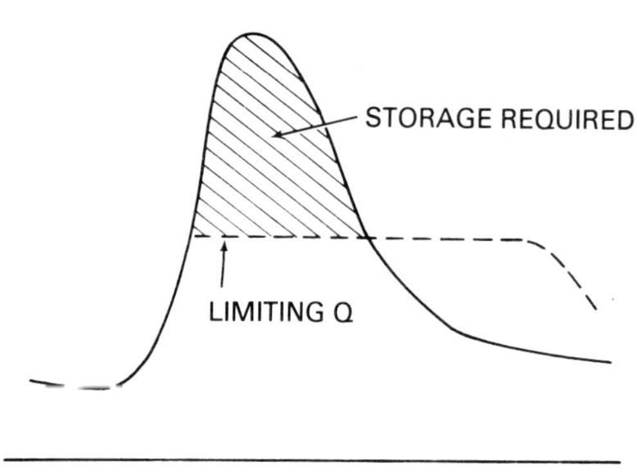

Fig. 11.7. Method of defining storage required to control a major flood.

Major floods are not always widely separated in time. If one large flood is closely followed by another, the storage needed to control the discharge may be larger than that defined by analysis of a single design hydrograph. The frequency mass-curve method provides the storage needed for such occurrences. From a streamflow record the highest mean discharges for various numbers of consecutive days in each year are extracted, and frequency curves prepared. The examples shown in Figure 11.8 were defined graphically because they are of various

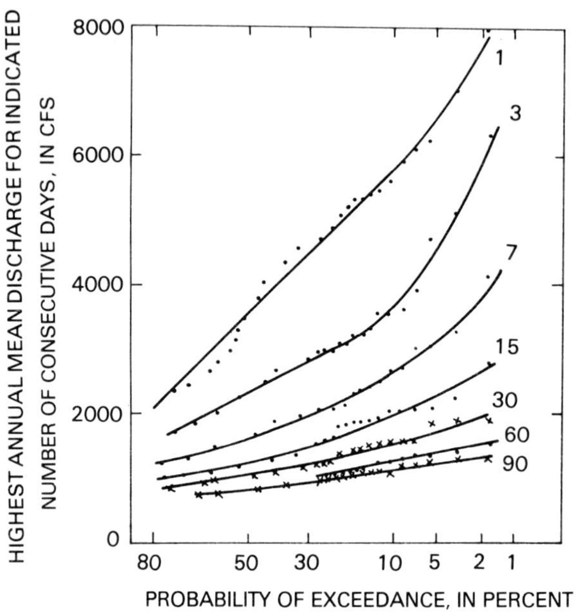

Fig. 11.8. Frequency curves of annual highest mean discharges for various numbers of consecutive days, Toccoa River, Georgia.

shapes and only the upper ends are needed. Discharges at selected probabilities are taken from these frequency curves and multiplied by the appropriate number of days to get volumes. These volumes are plotted against days to define a frequency mass curve such as that of Figure 11.9. The method of determining storages required to limit outflows to specified discharges is similar to that used with the conventional mass curve; the storage required is the maximum ordinate between the curve and a line representing the specified maximum discharge.

The amount of storage defined as above may not be adequate at times to limit the outflow to the specified level unless releases are near the specified level until the storage is depleted. Such sustained high releases are usually undesirable and need not be made under all conditions. Operation of a flood control reservoir usually is based on a "rule curve" which specifies the releases depending on reservoir stage, inflow forecast, and other relevant information.

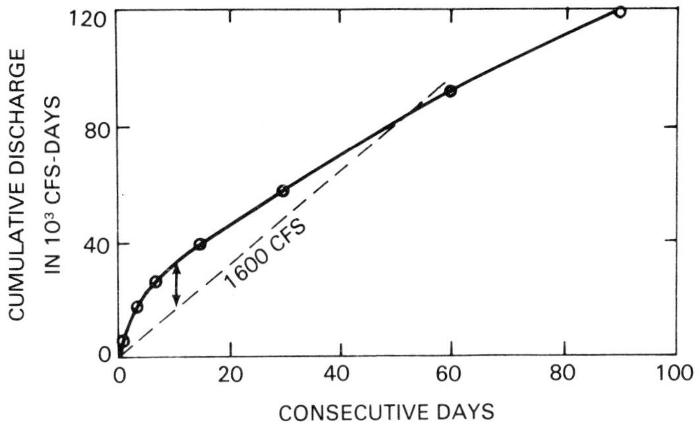

Fig. 11.9. Frequency mass curve of annual maximum flood volumes for 2 percent probability of deficiency. Data from Figure 11.8.

More sophisticated methods are used to design important flood-control projects particularly where several reservoirs are proposed. Various combinations of storage capacities and operating plans are evaluated using recorded or synthesized streamflows.

Small flood-control reservoirs commonly have fixed outlets through which the rate of release decreases with reservoir stage. The U.S. Soil Conservation Service used many such floodwater-retarding structures on small subwatersheds to control flood runoff from the major watershed. A typical structure and a description of the program in Texas is given by Gilbert and Sauer (1970).

11.3 DEPENDABLE FLOWS WITHOUT STORAGE

Many water-supply needs can be provided by the natural flow of a stream. The adequacy of a stream for a particular use may depend on whether a temporarily reduced supply can be tolerated. A municipal water system can provide less water than usual for limited periods without serious consequences to the majority of its customers. Hydropower plants and plants requiring process or cooling water can be shut down for short periods, but at some cost. But a nuclear power plant must have an assured supply of cooling water.

Maintenance of certain instream flows is desirable or legally required for preserving the aquatic habitat, for providing dilution for waste disposal, and for esthetics (American Fisheries Society, 1976). Realistic instream flows cannot be established without information on streamflow.

Annual low-flow frequency curves show the dependable flow without storage. The 7-day or 30-day curve is commonly used, at a probability of 0.1 or less according to the consequences of that discharge not being available. Seasonal dependable supplies can be determined from seasonal low-flow frequency curves.

11.4 BRIDGE AND CULVERT OPENINGS

The optimum size and geometry of a waterway opening depends principally on the design flood and the channel characteristics. Economic studies lead to selection of the appropriate design flood. Interstate highway crossings are designed to pass the 100-year flood. Minor road crossings where road overflow would result in little damage and/or inconvenience may be designed to pass a 5 or 10-year flood. The peak discharge for the selected recurrence interval at the usually ungaged site is determined as described in Chapter 8.

Design of the opening requires consideration of flow distribution in the approach channel, estimation of the stage-discharge relation at the crossing site, and computation of velocities through the proposed openings and of the resultant backwater (Bureau of Public Roads, 1960). Procedures for computing flows through bridges and culverts are given by Matthai (1968) and Bodhaine (1968) respectively. See Colson and Schneider (1983) for flows through multiple bridges.

Some culverts are designed to take advantage of the storage above the highway fill, thereby reducing the flood peak and the required culvert size. The reduction attained depends on the volume available for storage and on the volume of the flood hydrograph. At ungaged sites the hydrograph volume may be estimated from the flood peak. The relation

$$V = 0.131\ Q^{0.878}$$

where V is hydrograph volume in acre-feet and Q is peak discharge in cfs, was derived from hydrographs for 35 small Wyoming streams (Craig and Rankl, 1978). Peak discharge for the design recurrence interval can be estimated by one of the methods described in Chapter 8. Craig and Rankl also described a procedure for analyzing the reduction in peak flow and the increased head on the culvert for various culvert sizes and upstream topographies. Eychaner (1976) developed a relation of volume to peak for streams in southern Utah and a dimensionless graph showing the reduction in required structure capacity with storage (Fig. 11.10).

11.5 FORECASTING STREAMFLOW

A forecast uses information now available that will affect a future flow. For example the water content of a snowpack is information about runoff that will result from melting of the pack. On the other hand, a prediction is an inference regarding a future event based on probability theory.

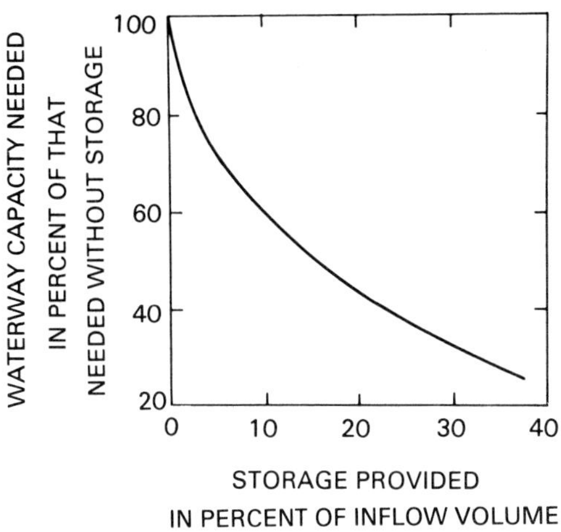

Fig. 11.10. Reduction in needed waterway opening by providing storage. For southern Utah. (After Eychaner, 1976).

Forecasts are commonly made of flood peaks or hydrographs, runoff volumes from snowmelt, and runoff for one or several months during a low-flow season. Streamflow and climatological data are needed to calibrate some forecasting models, and one or both types of data may be needed to make the forecast. Weather forecasts are valuable for making short-term streamflow forecasts such as for floods. Guidelines and examples of forecasting methods are given by WMO (1969a), NOAA (1972), and by ASCE (1980).

11.5.1 Floods

The purpose of forecasting floods is to provide information on flood discharges, stages, and times of occurrence in order to eliminate loss of life and reduce damage to property. Another purpose is to provide information for reservoir operation (Nordenson & Richards, 1964).

Data needed include known or forecast storm precipitation, current streamflow, and the stage-discharge relation at the site of interest. The volume of runoff from the storm is estimated and then is distributed in time to obtain the hydrograph. This may be done graphically but digital modeling is widely used. Nordenson (1969) described the application of conceptual catchment models to river forecasting, and Sittner (1976) reported results of a WMO project on intercomparison of conceptual models used in hydrological forecasting. Details of modeling are given by Overton and Meadows (1976).

Hydrographs derived from precipitation, or measured at a gaging station, may be routed downstream to other points of interest (Linsley, Kohler, and Paulhus, 1975, p. 287-315). A routing component is included in most conceptual models used for forecasting.

Discharge is forecast by most models but stage is the characteristic needed to assess the extent of flooding. Therefore a stage-discharge relation must be defined for transferring discharge to stage. If a rated gaging station is in the vicinity, its stage-discharge relation can be used although it may have to be extended well above the highest point of definition for application to unusual discharges.

The time between the precipitation and the peak discharge in a stream ranges from less than an hour to many days; it depends on the size of the stream and its runoff characteristics. To be of value the forecast must be made and disseminated quickly. Transmission of input data via satellite is most satisfactory.

Forecasts of small stream flood peaks cannot be made quickly enough following the storms to be useful. Consequently the National Weather Service issues flood warnings for small streams on the basis of forecasts of heavy rainfall.

11.5.2 Snowmelt runoff

The summer water supply for irrigation, hydropower, and other uses is derived from melting of the high mountain snowpack in many basins of western United States. Advance knowledge of the magnitude of runoff allows time to plan for the optimum use of that often inadequate resource. For example, forecasts as of April 1, 1934 showed Utah to be facing the worst drought in her history; prompt implementation of a conservation program minimized its effects (Clyde, 1951).

Records of streamflow, snow surveys, and precipitation for a number of concurrent years are used to develop relations for estimating runoff during the snowmelt period. Other data that may be useful include base flow of the stream previous to the beginning of the melt, and the condition of the soil underneath the snowpack - frozen or unfrozen, wet or dry. Precipitation during the runoff period helps to refine the relation but, of course, it cannot be estimated at time of forecast.

Regression models are widely used for forecasting seasonal runoff (Soil Conservation Service, 1972). Figure 11.11 shows a graphical relation of April through July runoff to maximum water content of the Spring snowpack.

Seasonal runoff from melting of a snowpack also may be estimated from precipitation records at lower elevations. In some basins, runoff estimates from precipitation are comparable in reliability to ones made from snow-survey data. For such basins the precipitation method would be used because of the negligible cost of low-elevation precipitation records relative to the cost of snow surveys. Kohler (1959) suggested that snow survey and precipitation data are complementary and that both can be used to advantage. Hawley, McCuen, and Rango (1980) evaluated 3 models for forecasting snowmelt runoff volumes and found the regression model most accurate for 60 days or longer.

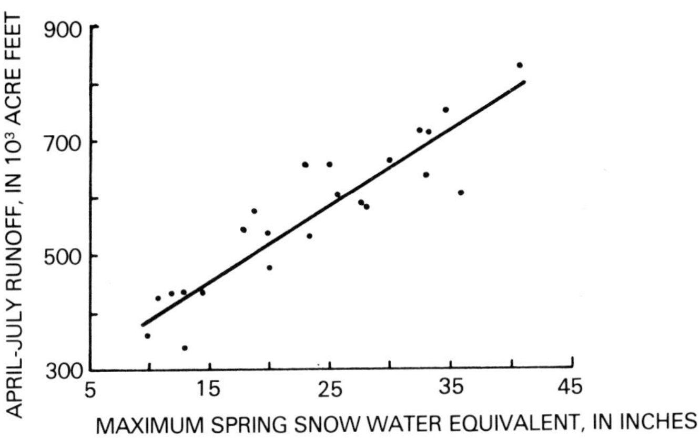

Fig. 11.11. Relation for forecasting snowmelt runoff of Rogue River, Oregon, from water content on Diamond Lake snow course (From Soil Conservation Service, 1972).

Forecasts of runoff from snowmelt are made monthly during late winter and spring. The best early estimate of seasonal runoff from a snowpack usually is the one based on snow survey data as of the date for which the snowpack water content is a maximum (Riggs, 1980).

Autumn runoff of some streams in semiarid regions can be forecast from the spring snowpack which provides the source of autumn runoff through recharge of ground water. A forecast relation of this type is shown in Figure 11.12.

Runoff from a snowpack is highly related to the temperature regime as indicated by the hydrograph of daily flow in Figure 11.13 for a small stream. And even for large streams the hydrograph for the snow runoff period depends on the temperature regime (and of course, on rainfall during the period). Thus reliable short-term forecasts of runoff from snowmelt are limited to a few days for which weather can be forecast. Schermerhorn (1961) used a single temperature index for making 3-day forecasts. His index is based on the unit-hydrograph principle and is constructed by weighting antecedent temperatures in proportion to the ordinates of the basin's snowmelt unitgraph. Zhidikov, et. al. (1976) developed a snowmelt runoff model for forecasting flows of relatively small rivers five days in advance. Model inputs are water equivalent of snow cover, precipitation, antecedent moisture content, daily snowmelt, non-uniformity of snow cover, retention capacity of the basin, and percent of forest cover.

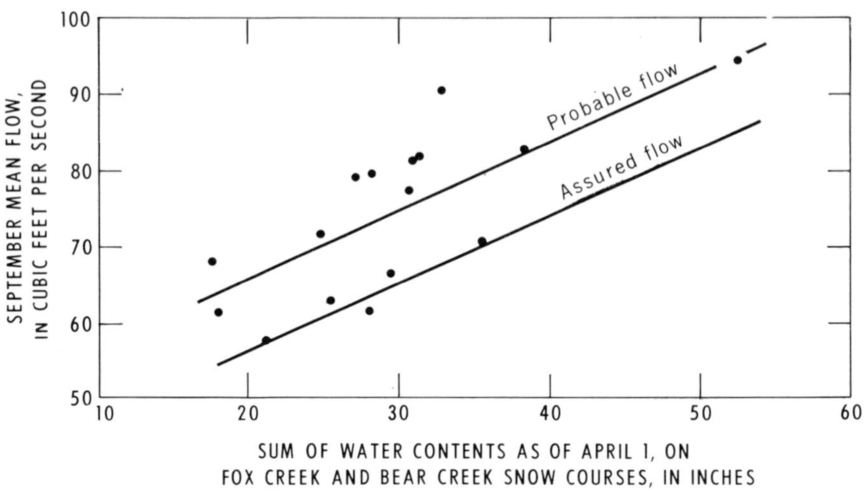

Fig. 11.12. Forecasting relations for Bruneau River, Idaho.

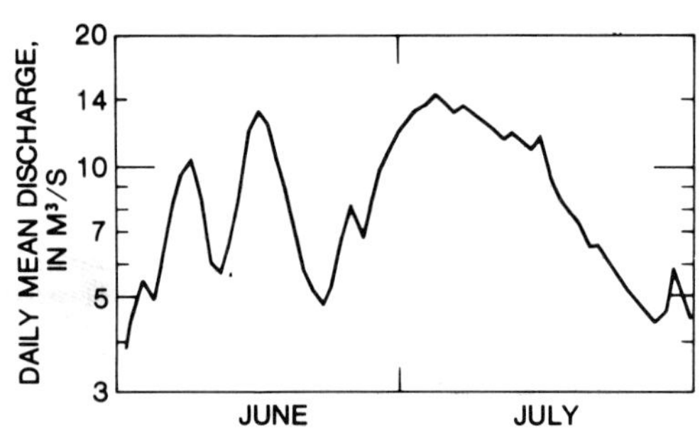

Fig. 11.13. Snowmelt runoff during a period of negligible precipitation. Gage at elevation 2700 m (Riggs, 1980).

Winter and early-spring runoff from basins of low relief provides needed water and may produce damaging floods. The depth of snow and the areal extent of snow cover on a basin can be analyzed with various assumed temperature and precipitation patterns to show the probability of various magnitudes of runoff volume and of peak discharge. WMO (1969b, p. 117-135) describes some methods of estimating the maximum floods from snowmelt and from a combination of snowmelt and rainfall, using an example from eastern Canada. Baker (1972) found a high correlation between the cold-period precipitation (snow or rain) and early-spring runoff in Minnesota where soils freeze annually. Results of his study indicate that this precipitation can be entirely accounted for by runoff and sublimation (or evaporation). The peak discharge depends, of course, on the temperature regime during the period of melt.

Snow-survey data were the basis for a timely warning of the disastrous flood of March 1936 on the Androscoggin River basin in Maine. The snow contained 250 mm of water at low elevations and up to 450 mm at higher elevations. The flood resulted from melting of the snowpack during a week in which 250 mm of rain fell on the basin (Church, 1949).

11.5.3 Seasonal low flows

Forecasts of monthly mean flows for one or more months are useful in managing a limited water supply for domestic, industrial, and hydropower uses.

The amount of water in the ground and in natural surface storage at the time of forecast is represented by the base flow of the stream. This base flow is the information used in forecasting monthly or seasonal flows. In general, the most reliable forecasts can be made on streams which have dependable base flows and receive little precipitation during the forecast period. Assured flow can be estimated by projecting the base-flow recession curve (see Chapter 5) to the end of the forecast period. In Figure 11.14 the recession curve is projected from the end of May when the discharge is nearly all base flow. If streamflow

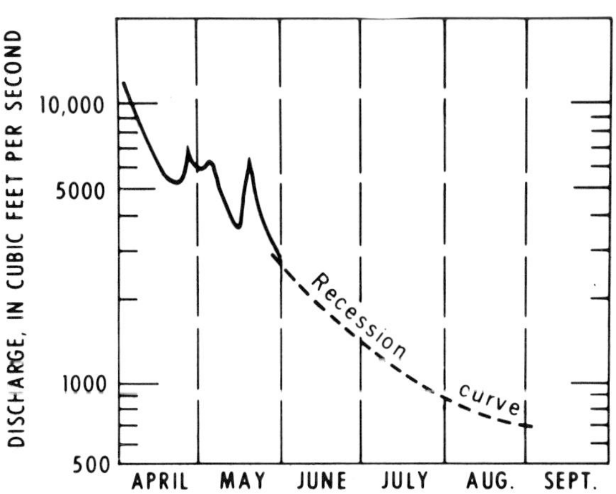

Fig. 11.14. Estimating assured flow by projecting the base-flow recession curve, Potomac River.

at time of forecast is not entirely base flow the amount of base flow must be estimated and the recession curve projected from that discharge rather than from the total flow. To demonstrate this, assume that the hydrograph of Figure 11.15 is known only through July. An estimate of the assured August base flow requires projection of the base-flow recession curve from early July to July 31, increasing the July 31 base flow to account for recharge during July, and projection of the recession curve through August. But the total assured flow includes water in channel storage on July 31; this is estimated by use of a direct-flow recession curve.

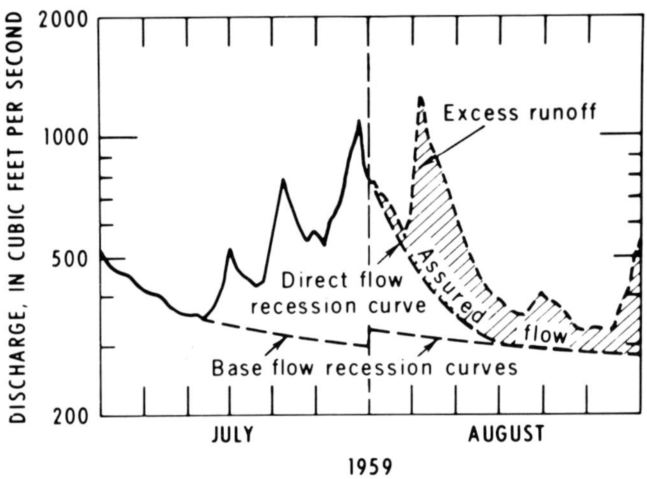

Fig. 11.15. Hydrograph for James River, Virginia showing the method of forecasting assured flow for August by projecting direct-flow and base-flow recession curves.

The assured flow will be the flow if there is no rain during the forecast period. Additional flow due to rain can be evaluated from a frequency curve of runoff in excess of assured; the frequency curve should be for the same month or period of forecast and can be defined by streamflow records at the site. The definition of excess is shown on Figure 11.15. The frequency curve based on many Augusts is in Figure 11.16.

Forecasts for various probabilities of exceedance can be obtained by adding to the assured flow discharges at selected probabilities. Details of the forecasting procedures, and applications, are given by Riggs (1953) and Riggs and Hanson (1969).

A statistical method of forecasting monthly flow is useful where flow for some future month or period depends to some extent on the water in storage at time of forecast. The flow for the forecast period is related to the known flow for an earlier period using streamflow records as shown in Figure 11.17. Interpretation of the plotted points depends on the answer sought; the line on Figure 11.17 may be used to estimate the assured flow; a line averaging the points would give the probable flow.

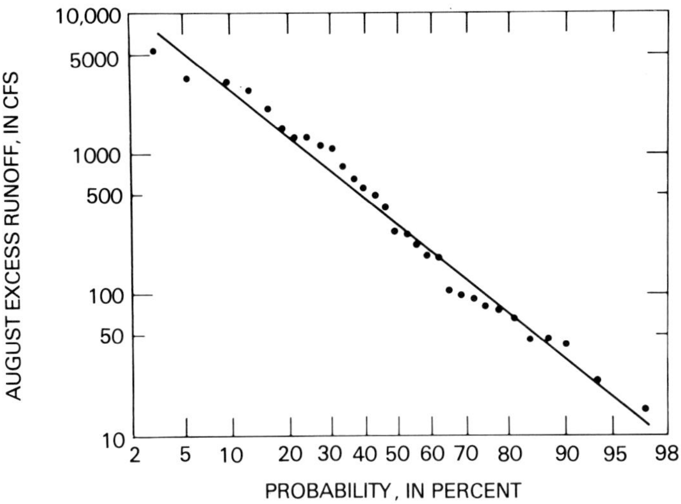

Fig. 11.16. Frequency curve of August excess runoff as defined by Figure 11.15.

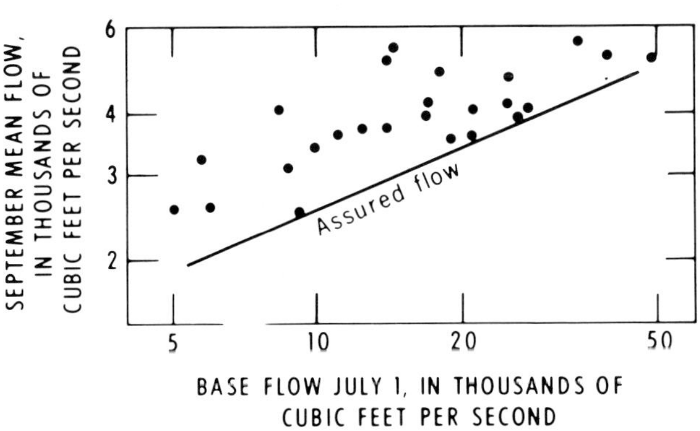

Fig. 11.17. Forecasting relation for Salmon River at Whitebird, Idaho.

11.6 STREAMFLOW DROUGHTS

A drought is broadly defined as a deficiency in precipitation that affects man's activities and interests. The effects depend on the severity, duration, and geographical extent of the precipitation deficiency, and on whether precipitation is used directly, for example to maintain soil moisture, or whether water supplies are drawn from streams or from ground water.

In non-irrigated agriculture, lack of rain for a few weeks during the growing season will reduce the yield or destroy the crop. Likewise, residential or municipal water supplies which depend on runoff from roofs or paved areas and for which limited storage is available, will be inadequate if not replenished frequently. Stock water in semiarid regions is often provided by reservoirs, called stock tanks, on small ephemeral streams. Maintenance of water in these tanks requires appreciable periodic rainfall. Users of water for these purposes would consider a relatively short period without rainfall as a drought.

A lack of rainfall for several consecutive summer months will not constitute a drought if the normal for those months is near zero, as it is in San Francisco, because a drought is a deficiency from normal.

Lack of rain for a few weeks or a month may have no appreciable effect on water supplies derived from streamflow or from ground water. But lack of rain for an extended period might result in streamflow dropping below the demand; river flow recedes at a characteristic rate during periods of no precipitation as the ground water reservoir is drained. For example, a couple of summer months with no storms large enough to produce runoff might let the Potomac River flow recede to less than the amount required for municipal supply of Washington, D.C. The effects of rain depend on its magnitude and intensity. Fairly intense rainfall is needed to produce runoff and thus to augment streamflow. Frequent light rains may be adequate to support vegetation but may not produce any runoff or recharge.

Although streamflow drought is caused by lack of normal precipitation, the effects of precipitation deficiencies are delayed because of the influence of the land on the disposition of precipitation.

Streamflow during the summer period also may be affected by precipitation in the previous winter and spring when the ground water body, which sustains streamflow during dry weather, is recharged. If the recharge is large, the base flow of the stream will be large at the beginning of the summer and will be less likely to be adversely affected by deficient summer rainfall.

Where water supplies are drawn from surface and ground-water reservoirs, a lack of rainfall for a few weeks, or even a substantial deficit for the whole year, may have little effect if the reservoirs are of adequate size. The critical drought under such conditions is caused by deficient precipitation for several successive years so that the usable storage, both surface and underground, becomes progressively depleted until the usual amount of withdrawals cannot be made. The 1976-77 drought in California is an example. The Sierra Nevada snowpack, the major source of supply, was lowest of record at many sites in 1976, and was even lower in 1977. As a consequence, surface reservoir storage was depleted, streamflow was inadequate for irrigation and other supplies, the reduced flow into San Francisco Bay allowed salt water to move upstream in the Delta, and the ground-water table was lowered up to 30 feet by pumping to augment the reduced supplies from the usual sources (Matthai, 1979).

Other notable United States droughts are those of the 1930's (Hoyt, 1936, 1938), and 1950's (Gatewood et al, 1964; Nace and Pluhowski, 1965), and the 1960's (Barksdale et al, 1966). Meteorological aspects can be described by the Palmer (1965) index.

11.6.1 Seasonal streamflow drought

On an unregulated stream the usable water supply depends on the magnitude of the seasonal low flow which varies from year to year. Streams with substantial base flows recede slowly, permitting a reasonable estimate of the base flow a month or months in the future in the absence of precipitation (see Section 11.5.3). Streams with little base flow recede rapidly and are not sources of dependable supplies unless rainfall occurs regularly.

The severity of a seasonal drought can be expressed by the recurrence interval of the 30-day (or other) annual minimum flow. This information is available, of course, only after the drought is over. Figure 11.18 shows significant annual droughts on French Broad River, North Carolina.

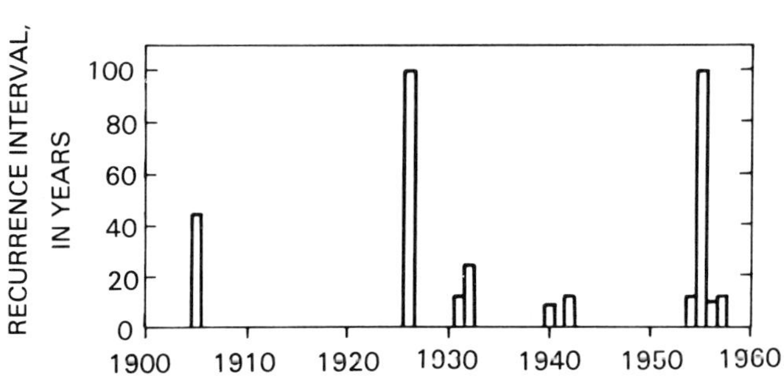

Fig. 11.18. Significant annual 30-day low flows, French Broad River, North Carolina.

11.6.2 Multiyear streamflow drought

When annual mean streamflows are below normal for two or more consecutive years, the draft available from storage during that period may be greatly reduced even though none of the seasonal shortages for individual years (on streams without storage) are large. The ground-water supply to the stream also may be progressively depleted because of reduced recharge. Furthermore the effect of a multiyear drought on draft available from storage depends on the size of the storage reservoir, so that the same precipitation deficiency might appear to produce different effects on adjacent strams if those streams did not have the same storage relative to their mean flows.

The major droughts throughout the world have lasted for several years. During the drought of the 1930's in United States much farmland was abandoned in the Great Plains because of lack of rain and the consequent wind erosion. The magnitude and duration of that drought can be seen from the streamflow record of Red River of the North in the Dakotas and Minnesota (Fig. 11.19); the annual mean flows were below average from 1928 to 1941 and the flows for 8 of those years were lower than any prior to 1928 as far back as 1884.

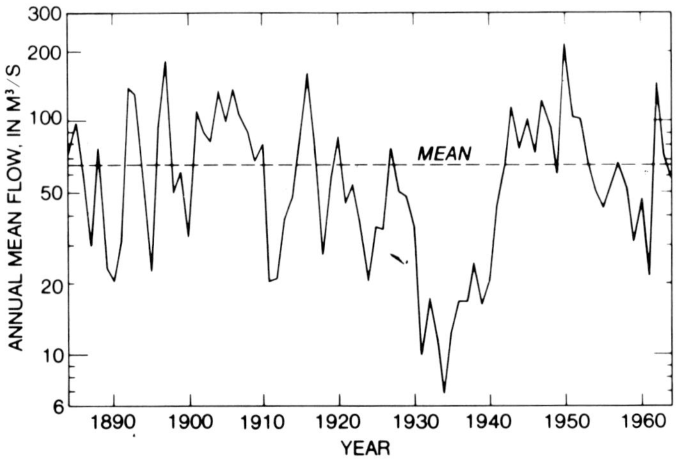

Fig. 11.19. Annual mean flows of Red River, North Dakota-Minnesota showing drought of the 1930s.

Multiyear droughts cannot be simply described because they have 3 characteristics: magnitude, duration, and areal extent. Riggs (1979) proposed a method of quantifying multiyear drought severity for use in comparing droughts. The indices are based on the draft that could have been obtained through the drought period under the assumption that a storage equivalent to the mean annual runoff existed in the basin. This approach makes the indices consistent from time to time at a site and consistent among sites having the same coefficient of variation of annual means. The procedure for defining the drought index is as follows:

1. Plot hydrograph of annual means for an unregulated stream.
2. Identify drought periods from hydrograph.
3. Plot a mass curve of annual means encompassing each drought period.
4. On each mass curve draw a draft line limited by a storage equal to the mean annual runoff volume and by a period of deficient flow of 6 years.
5. Compute draft rate as percent of mean flow.
6. Compute drought index as the product of 100 minus draft rate and the square root of the deficient period in years.

The procedure is illustrated in Figure 11.20 using flow data from Figure 11.19. Results for Red River and another river are plotted in Figure 11.21.

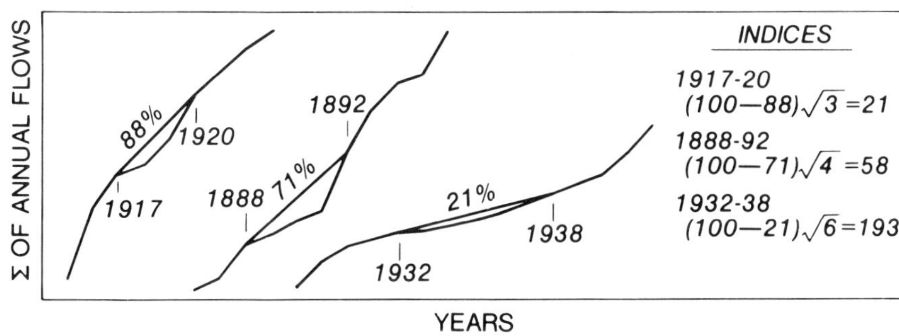

Fig. 11.20. Analysis of available draft and computation of drought indices, Red River, North Dakota - Minnesota.

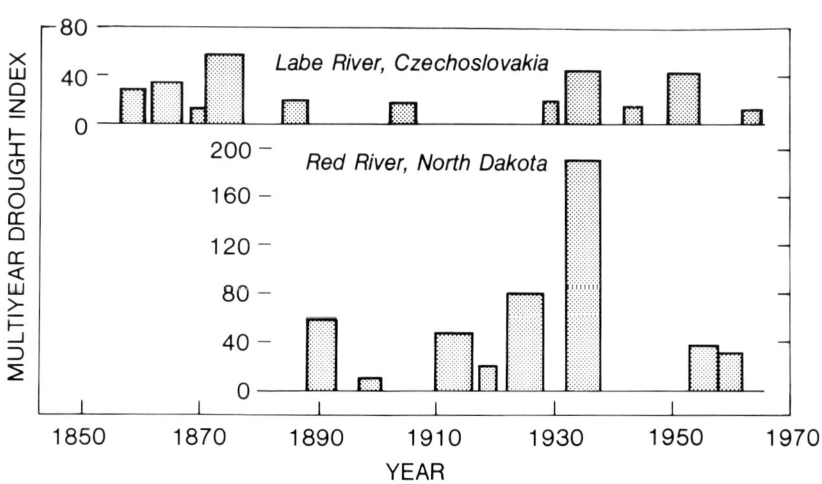

Fig. 11.21. Multiyear drought indices for 2 rivers. Duration of drought is shown by width of block.

The above indices apply to individual streams. Regional indices can be derived by averaging the annual flows (in percent of their mean) of representative streams in the region and defining the indices as for a single stream. Figure 11.22 compares regional multiyear and seasonal drought indices for the Tennessee-Cumberland basins, USA.

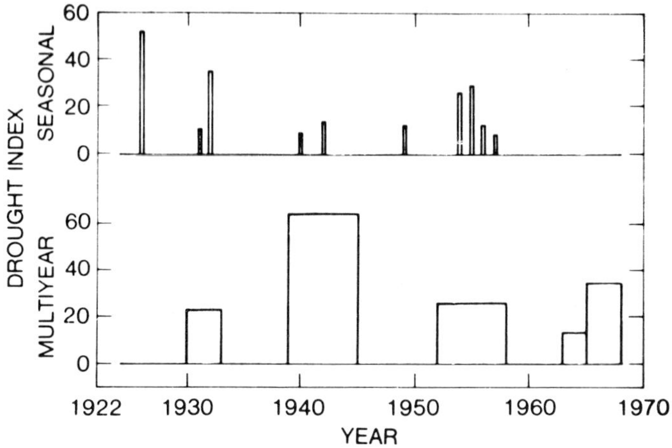

Fig. 11.22. Indices of seasonal and multiyear droughts in the Tennessee-Cumberland basins.

11.6.3 Adaptation to drought

The beginning of a hydrologic drought usually is not recognizable at the time. Periods of below normal flow are commonly terminated by precipitation within a few weeks or a month. Precipitation cannot be forecast reliably far enough in the future to indicate whether a moderate drought period is the beginning of a severe drought or whether the period will be terminated soon. Thus, adaptation to a drought usually involves actions that can be taken quickly when the severity is recognized. These include tapping an alternative source, transferring water from a low priority use to a higher one, reducing water use, and reusing water. Matthai (1979, p. 19-23) describes the changes in water use made during the 1976-77 drought in United States.

Forecasts of some impending annual droughts may be feasible. Estimates of runoff from the spring snowpack and estimates of summer low flow from spring precipitation are reliable on some streams. These forecasts provide more time to adapt to a deficient supply. During a multiyear drought short-term forecasts may be feasible but ones for half a year or more are not. Probability estimates may be used for planning and these should incorporate present reservoir storage and any restrictions in use that will be in effect (Hirsch, 1981).

Protection from severe water shortages can be provided only if the natural fluctuations in water supply are accurately known and are incorporated in the project design. If the regional supply is fully used in most years there will be no alternate temporary supply to fall back on during a drought.

11.7 FLOOD-PRONE AREA MAPPING

Definition of flood-prone areas is needed for flood plain management, a means of reducing flood losses by adapting land use to the flood hazard. A principal use of flood inundation maps is the setting of rates for flood insurance. The National Flood Insurance Act of 1968 and the Flood Disaster Protection Act of 1973 were enacted to inform communities about their flood dangers, to provide new construction reasonable protection from flood damage by prudent management of the flood plain, and to protect residents of flood-plain areas against financial loss from flooding. Waananen and others (1977) describe how knowledge of flood-prone areas affected land-use planning in the San Francisco Bay area.

A flood-prone area is that area which will be inundated by a flood of some specified recurrence interval, usually 100 years. Examples of flood-prone area maps are shown in Figures 11.23 and 11.24. The area inundated by the 100-year

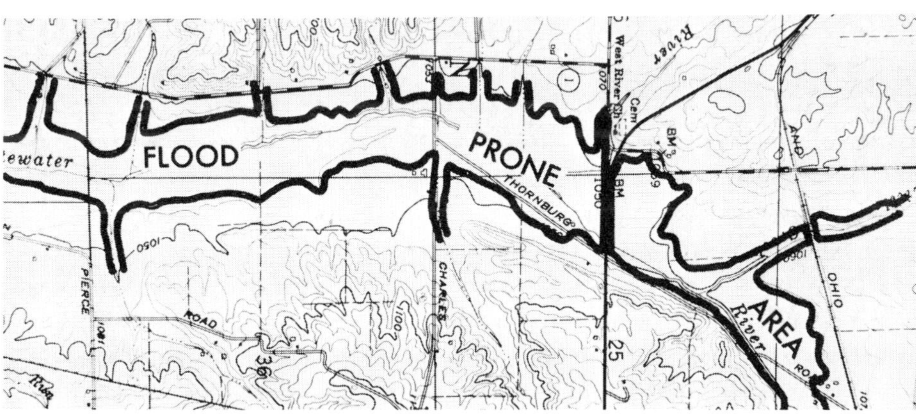

Fig. 11.23. Flood-prone area defined by an approximate method.

flood in Figure 11.23 was defined by an approximate method in which stream depth of the 100-year flood is estimated from a regional relation with drainage area or with mean annual flood discharge. In regions where flood characteristics, or channel shapes, are not closely related to drainage area, the mean annual flood discharge can be estimated by one of the methods given in Chapter 8. The estimated depth is used to delineate the inundated area on a topographic map. The reliability of the estimate of the inundated area can be improved if a profile of an observed flood, or an aerial photograph of flooding, is available for the reach.

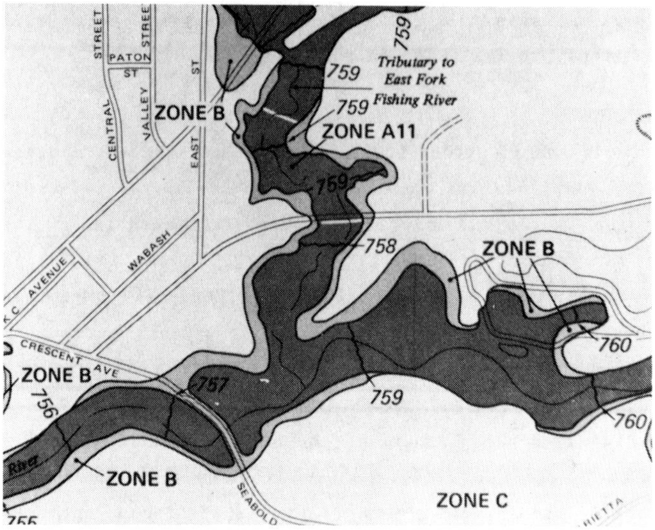

Fig. 11.24. Flood-prone area defined by step-backwater analysis.

More reliable, and much more expensive, results can be obtained by defining the water-surface profile through the reach by standard step-backwater analysis. This requires measured channel cross sections at intervals through the reach and estimates of channel roughness (Davidian, 1984). Computer programs are available for the analysis. The map of Figure 11.24 was prepared by this method.

Encroachment on flood plains, such as by artificial fill, usually reduces the flood-carrying capacity and increases flood heights, thus increasing flood hazards in areas beyond the encroachment itself. One aspect of flood-plain management involves balancing the economic gain from flood-plain development against the resulting increase in flood hazard. Under the floodway concept the area inundated by the 100-year flood is divided into a proposed floodway and floodway fringes. The floodway is the channel of the stream plus any adjacent flood-plain areas that must be kept free of encroachment in order that the 100-year flood be carried without a substantial increase in flood height (Fig. 11.25). The floodway fringes may be completely obstructed.

11.8 ESTIMATION OF ENVIRONMENTAL IMPACT

The effect on the water resource of some proposed change in a drainage basin depends on the flow characteristics at the present time. These characteristics usually must be estimated from regional relations, possibly supplemented by a few field measurements. Determination of the changes in response to the anticipated stress calls for an understanding of the hydrology, the knowledge of changes that occurred in other basins, and sound judgment to adjust for the different conditions.

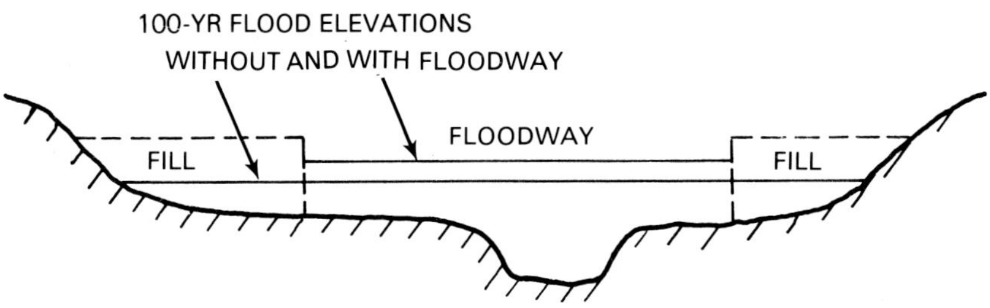

Fig. 11.25. Channel cross section showing floodway.

REFERENCES

American Fisheries Society, 1976, Instream flow needs; Proc. of Symposium by Am. Fisheries Society and Am. Soc. of Civil Engineers at Boise, Idaho, May 1976: American Fisheries Society, Bethesda, Md., 2 Vol.

ASCE, 1980, Proceedings of the Engineering Foundation Conference on Improved Hydrologic Forecasting, Why and How: New York, Am. Soc. of Civil Engineers.

Baker, D.G., 1972, Prediction of spring runoff; Water Resources Research, Vol. 8, No. 4, p. 966-972.

Banks, H.O., Chairman, 1964, Hydrology of spillway design: Jour. of Hydraulics Division, ASCE, Vol. 90, No. HY3, p. 235-310.

Barksdale, H.C., O'Brien, D., and Schneider, W.J., 1966, Effect of drought on water resources in the Northeast: U.S. Geol. Survey Hydrologic Atlas 243.

Bodhaine, G.L., 1968, Measurement of peak discharge at culverts by indirect methods: U.S. Geol. Survey Techniques of Water-Resources Investigations, Book 3, Chapt. A3, 60 p.

Bureau of Public Roads, 1960, Hydraulics of bridge waterways: U.S. Dept. Commerce, Bureau of Public Roads Hydraulic Design Series 1, 53 p.

Burges, S.J. and Linsley, R.K., 1971, Some factors influencing required reservoir storage: Jour. of Hydraulics Division, ASCE, Vol. 97, No. HY7, p. 977-991.

Church, J.E., 1949, Snow and snow surveying: Ice, in Meinzer, O.E., Hydrology: New York, Dover Publ., Inc.

Clyde, G.D., 1951, Benefits of snow surveying: Western Snow Conf. Proc., Nineteenth Annual Meeting, p. 84-89.

Colson, B.E. and Schneider, V.R., 1983, Backwater and discharge at highway crossings with multiple bridges in Louisiana and Mississippi: U.S. Geol. Survey Water-Resources Investigations Rept. 83-4065, 39 p.

Craig, G.S., Jr. and Rankl, J.G., 1978, Analysis of runoff from small drainage basins in Wyoming: U.S. Geol. Survey Water-Supply Paper 2056, 70 p.

Davidian, J., 1984, Computation of water-surface profiles: U.S. Geol. Survey Techniques of Water-Resources Investigations, Book 3, Chapt. D15.

Eychaner, J.H., 1976, Estimating runoff volumes and flood hydrographs in the Colorado River basin, southern Utah: U.S. Geol. Survey Water-Resources Investigations 76-102, 18 p.

Gatewood, J.S., Wilson, A., Thomas, H.E., and Kister, L.R., 1964, General effects of drought on water resources of the Southwest: U.S. Geol. Survey Prof. Paper 372-B, p. 1-55.

Gilbert, C.R. and Sauer, S.P., 1970, Hydrologic effects of floodwater retarding structures on Garza-Little Elm Reservoir, Texas: U.S. Geol. Survey Water-Supply Paper 1984, 95 p.

Hardison, C.H. and Furness, L.W., 1963, Discussion of paper by John B. Stall, "Reservoir mass analysis by a low flow series: Jour. of Sanitary Engineering Division," ASCE, Vol. 89, No. SA2, p. 119-122.

Hawley, M.E., McCuen, R.H., and Rango, A., 1980, Comparison of models for forecasting snowmelt runoff volumes: Water Resources Bulletin, Vol. 16, No. 5, October 1980, p.

Hershfield, D.M., 1961, Estimating probable maximum precipitation: Jour. of Hydraulics Division, ASCE, Vol. 87, p. 99-106.

Hirsch, R.M., 1981, Estimating probabilities of reservoir storage for the upper Delaware River basin: U.S. Geol. Survey Open-File Report 81-478.

Hoyt, J.C., 1936, Droughts of 1930-34: U.S. Geol. Survey Water-Supply Paper 680, 106 p.

Hoyt, J.C., 1938, Drought of 1936: U.S. Geol. Survey Water-Supply Paper 820, 62 p.

Kohler, M.A., 1959, Preliminary report on evaluating the utility of water-supply forecasts: Western Snow Conf. Proc., 17th Annual Meeting, April 1959, p. 26-33.

Langbein, W.B., 1958, Queuing theory and water storage: Jour. of Hydraulics Division, ASCE, Vol. 84, No. HY5, p. 1811-1 to 1811-24.

Leopold, L.B., 1959, Probability analysis applied to a water-supply problem: U.S. Geol. Survey Circ. 410, 18 p.

Linsley, R.K., Kohler, M.A., and Paulhus, J.L.H., 1975, Hydrology for engineers: New York, McGraw-Hill, 482 p.

Lof, G.O.G. and Hardison, C.H., 1966, Storage requirements for water in the United States: Water Resources Research, Vol. 2, No. 3, p. 323-354.

Matthai, H.F., 1968, Measurement of peak discharge at width contractions by indirect methods: U.S. Geol. Survey Techniques of Water-Resources Investigations, Book 3, Chapter A4, 44 p.

Matthai, H.F., 1979, Hydrologic and human aspects of the 1976-77 drought: U.S. Geol. Survey Prof. Paper 1130, 84 p.

Nace, R.L. and Pluhowski, E.J., 1965, Drought of the 1950s with special reference to the midcontinent: U.S. Geol. Survey Water-Supply Paper 1804, 18 p.

NOAA, 1972, National Weather Service River Forecast System forecast procedures: U.S. Dept. Commerce; NOAA TM NWS HYDRO-14.

Nordenson, T.J. and Richards, M.M., 1964, River forecasting in Chow, V.T., ed., Handbook of Applied Hydrology: New York, McGraw-Hill, p. 25-98 to 25-111.

Nordenson, T.J., 1969, The application of conceptual catchment models to river forecasting: Hydrological Forecasting, World Meteorological Organization Technical Note 92 (WMO-No. 228. TP. 122), p. 181-192.

Overton, D.E. and Meadows, M.E., 1976, Stormwater modeling: New York, Academic Press, 358 p.

Palmer, W.C., 1965, Meteorological drought: U.S. Dept. Commerce Weather Bureau Research Paper No. 45, 58 p.

Riggs, H.C., 1953, A method of forecasting low flow of streams: Trans. Am. Geophysical Union, Vol. 34, No. 3, p. 427-434.

Riggs, H.C., 1979, Characterizing streamflow droughts: Hydrological Aspects of Droughts, International Symposium, 3-7 December 1979, New Delhi, Proc. Vol. 1, p. 331-338.

Riggs, H.C., 1980, Runoff estimates from snowmelt in Pollution and Water Resources, Columbia University Seminar Series, Vol. XIII, Part 2: New York, Pergamon Press, p. 71-80.

Riggs, H.C. and Hanson, R.L., 1969, Seasonal low-flow forecasting: World Meteorological Organization Tech. Note No. 92, WMO-No. 228, TP 122, Hydrological Forecasting, Geneva, Switzerland, p. 286-289.

Riggs, H.C. and Hardison, C.H., 1973, Storage analyses for water supply: U.S. Geol. Survey Techniques of Water-Resources Investigations, Book 4, Chapter B2, 20 p.

Schermerhorn, V.P., 1961, Short-range snowmelt forecasts: Bull. of International Assoc. of Scientific Hydrology 6, No. 4, p. 75-82.

Sittner, W.T., 1976, WMO project on intercomparison of conceptual models used in hydrological forecasting: Hydrological Sciences Bull., Vol. 21, No. 1, p. 203-213.

Skelton, J., 1971, Carry over storage requirements for reservoir storage in Missouri: Missouri Geological Survey and Water Resources, Water Resources Rept. 27, 56 p.

Soil Conservation Service, 1972, Snow survey and water-supply forecasting: SCS National Engineering Handbook, Section 22, USDA, Soil Conservation Service.

Thomas, H.A., Jr. and Fiering, M.B., 1962, Mathematical synthesis of streamflow sequences for the analysis of river basins by simulation, in Maas and others, Design of Water Resource Systems: Cambridge, Mass., Harvard University Press.

Vanoni, V.A., Ed., 1975, Sedimentation engineering: ASCE Manual of Engineering Practice, 745 p.

Waananen, A.O., Limerinos, J.T., Kockelman, W.J., Spangle, W.E., and Blair, M.L., 1977, Flood-prone areas and land-use planning - selected examples from the San Francisco Bay area, California: U.S. Geol. Survey Prof. Paper 942, 75 p.

WMO, 1969a, Hydrological forecasting: World Meteorological Organization, Tech. Note No. 92, WMO-No. 228 TP. 122, Geneva, Switzerland.

WMO, 1969b, Estimation of maximum floods: World Meteorological Organization Tech. Note No. 98, WMO-No. 233, TP. 126.

WMO, 1973, Manual for estimation of probable maximum precipitation: World Meteorological Organization Operational Hydrology Rept. 1, WMO-No. 332, Geneva, Switzerland.

Zhidikov, A.P., Levin, A.G., Nechaeva, N.S., and Popov, E.G., 1976, A snowmelt runoff model and its application of short range forecasting of flood discharges: Hydrological Sciences Bull. 21, No. 1, p. 195-202.

Chapter 12

SOURCES OF DATA AND INFORMATION

12.1 INTRODUCTION

In the United States most of the sites for which water data are available are listed in the Master Water Data Index of NAWDEX (Edwards, 1979). The Index provides for each site the geographic location, the data-collecting organization, the types of data available, the periods of time for which the data are available, the major water-data parameters for which data are available, the frequency of measurement of the parameters, and the media in which the data are stored. Local Assistance Centers of NAWDEX can assist users in locating data for specific streams or areas. The user may then obtain the data from the organizations or media where stored.

NAWDEX is affiliated with the Water Resources Document Reference Center (WATDOC) of the Inland Waters Directorate, Canadian Department of Fisheries and The Environment, and with data services of several State-governmental organizations.

Further information on NAWDEX services can be obtained from

National Water Data Exchange
U.S. Geological Survey
421 National Center
Reston, Virginia 22092

In countries other than United States, the available water-resource data may be obtained from those governmental agencies whose work requires such data.

Understanding and analysis of water-resource data requires additional information, much of which can be extracted from climatic records and maps. Interpreted data, and previous hydrologic analyses, are particularly helpful. The principal sources of these types of information in the United States are included in this chapter.

12.2 WATER-RESOURCES DATA

The principal source of water-resource data in the United States is the U.S. Geological Survey whose data are published in annual data reports on a State-boundary basis. These reports include records of stage, discharge, and water quality (including sediment transport) of streams; stage and contents of lakes and reservoirs; and water levels and water quality of ground-water in wells. Previous to 1961 such data were published in Water-Supply Papers. Compilations of monthly and annual streamflows and of annual maximum and minimum flows through 1960 are published in Water-Supply Papers 1301-1319, 1372 (through 1950) and 1721-1740 (1951-60).

Data of record also are stored in Reston, Virginia in the computerized system called WATSTORE. Files are maintained for the storage of (1) surface-water, quality-of-water, and ground-water data measured on a daily or a continuous basis, (2) annual peak stages and discharges for streamflow stations, (3) chemical analyses for surface-water and ground-water sites, (4) water-data parameters measured more frequently than daily, (5) geologic and inventory data for ground-water sites, and (6) summary data on water use.

These data can be retrieved from WATSTORE in computer-printed tables or in conjunction with programs for the statistical analysis of certain of those data. For example, programs for defining and plotting frequency curves and for making regression analyses can utilize direct retrieval from WATSTORE. Information about the availability of specific types of data, the aquisition of data or products, and user charges can be obtained from local offices of the Geological Survey (Kilpatrick, 1981).

Water-quality parameters collected by several Federal organizations are in the STORET system of the U.S. Environmental Protection Agency. Data in STORET can be retrieved through the USGS NAWDEX system.

Surface Water Reference Index, Canada 1981, lists most of the Canadian stations for which streamflow data are available and refers to other publications for additional ones. Streamflows and related records on individual stations may be obtained from the Regional Offices listed or from Director, Water Resources Branch, Department of the Environment, Ottawa, Ontario.

Some streamflow data for various parts of the world are in publications of the World Meteorological Organization, Geneva, Switzerland.

12.3 CLIMATIC DATA

The Climatic Atlas of the United States (ESSA, 1968) depicts the climate of the United States in terms of the distribution and variation of constituent climatic elements. Principal subjects are temperature means and ranges by months, normal total monthly and annual precipitation, and pan and lake evaporation. Other subjects are snowfall, wind, sunshine, relative humidity and additional characteristics of temperature and precipitation.

Detailed climatic information is available from

NOAA, NCDC
National Climatic Center
Federal Building
Asheville, North Carolina 28801

which can supply air temperature, dew point, heating and cooling degree days, rainfall, snowfall, barometric pressure, wind direction and speed, solar radiation, and sky cover for about 300 NWS stations nationwide. Also available are daily maximum and minimum temperatures and total daily precipitation for approximately 12,000 sites. Monthly and yearly summaries of precipitation are also available.

Average annual lake and Class A pan evaporation in inches are shown on maps of the conterminous United States (Farnsworth and others, 1982). Monthly, seasonal, and annual averages of estimated pan evaporation are compiled by Farnsworth and Thompson (1982).

12.4 MAPS

The National Cartographic Information Center (NCIC), U.S. Geological Survey, Reston, Va. 22092, maintains a computer index of cartographic products of USGS and other federal agencies. These include topographic, geologic, and land-use maps, aerial photographs, and satellite imagery. Although NCIC is primarily the source of information, some products are available from the Regional Mapping Center Units (USGS, 1981). Printed indexes to topographic map coverage by States are available. Although USGS is the major source, maps are also available from other federal, state, and local agencies, and from commercial dealers.

Soil maps showing infiltration characteristics are published by the Soil Conservation Service of the U.S. Department of Agriculture and where available, may be obtained from the local Soil Conservation Service offices.

12.5 CURRENT CONDITIONS AND OUTLOOKS

Streamflow data are needed on a "real-time" basis, that is within an hour or less after its occurrence, for flood forecasting and for certain water-management decisions. Data from a gaging station can be transmitted over telephone lines, by radio, or by satellite. See Transmission of Hydrologic Data in Chapter 3. Data from many streams are being transmitted daily, or more frequently as required. Several hundred sites are reporting (1981) via satellite. The District Offices of USGS can provide the names of the sites and the type, frequency, and method of data transmittal.

National Water Conditions (formerly Water Resources Review), published shortly after the end of each month by USGS, describes the streamflow and ground-water conditions in United States and southern Canada during the month in relation to normal. It also tabulates the usable contents of selected reservoirs, the flows of large rivers, and the water quality of a few large rivers. Descriptions of significant floods and droughts are included.

Weekly Weather and Crop Bulletin is prepared jointly by U.S. Department of Commerce (NOAA, National Weather Service) and U.S. Department of Agriculture. Principal contents are national weather summary, weather data for selected cities, State summaries of weather and agriculture, and international weather and crop summaries.

Water Supply Outlook for the Western United States is published jointly by National Weather Service (NOAA) and Soil Conservation Service (USDA) following the principal snow survey dates from January 1 to May 1. Snow survey data are included. These reports also contain references to other water supply outlooks for the State of California and for the western Canadian Provinces.

Average Monthly Weather Outlook, by NOAA, National Weather Service, gives mapped estimates of whether the average rainfall for the next 30 days will be heavier or lighter than normal; these are not forecasts in the meteorological sense. Similarly, the outlook for temperature is mapped in terms of its being below, above, or near normal. These outlooks are principally for North America but some information is given for Europe, North Africa, and Asia.

12.6 INTERPRETED DATA

Mean runoff in United States has been mapped by Busby (1966); his map is also published in USGS (1970). For yearly variations in runoff see Busby (1963).

Flood frequency curves for all gaged sites in the United States that had more than 10 years of recorded natural (unregulated) streamflow were prepared in 1977; some of these have been updated. They are available for inspection at the District Offices of the Water Resources Division of USGS.

Storm-rainfall frequencies in the conterminous United States for durations from 1/2 to 24 hours and for recurrence intervals up to 100 years have been defined and mapped by Weather Bureau (1961) and NOAA (1973).

Probable maximum precipitation in eastern United States has been estimated by the National Weather Service. See Schreiner and Riedel (1978) and Ho and Riedel (1980).

12.7 GENERAL

A bibliographic data service that relates specifically to water resources activities is provided by the Water Resources Scientific Information Center (WRSIC) of the U.S. Geological Survey. In 1979 the system had over 100,000 computerized abstracts that relate to water-resource subjects. Requests to WRSIC may be made through NAWDEX.

Water use data as of 1975 with projections to 2000 are given by WRC (1978) for 106 subregions of the United States. Solley and others (1983) have tabulated by States and by the 21 WRC regions water uses as of 1980 for public supply, rural, irrigation, self-supplied industrial, thermoelectric, and hydropower uses. Separate accountings are made for fresh and saline water withdrawals of both ground and surface water.

USGS (1983) reviews current hydrologic conditions and recent events in the United States and provides a broad overview of hydrologic issues facing the nation. The summary also describes water issues for each State.

A variety of water-resources data, facts, and statistics covering the subjects of climates, hydrology, surface and ground-water resources, water use and needs, and water quality has been compiled by Todd (1970).

REFERENCES

Busby, M.W., 1963, Yearly variations in runoff for the conterminous United States: U.S. Geol. Survey Water-Supply Paper 1669-S, 49 p.

Busby, M.W., 1966, Annual runoff in the conterminous United States: U.S. Geol. Survey Hydrologic Atlas 212, scale 1 to 7.5 million.

Edwards M.D., 1979, NAWDEX:A key to finding water data: U.S. Geol. Survey General Interest Publication, 15 p.

ESSA, 1968, Climatic Atlas of the United States: U.S. Dept. of Commerce, Environmental Science Services Admin., Environmental Data Service.

Farnsworth, R.K. and others, 1982, Evaporation atlas for the contiguous 48 United States: National Oceanic and Atmospheric Administration Tech. Rept. NWS 33, 26 p.

Farnsworth, R.K., and Thompson, E.S., 1982, Monthly, seasonal, and annual pan evaporation for the United States: National Oceanic and Atmospheric Administration Tech. Rept. NWS 34, 82 p.

Ho, F.P. and Riedel, J.T., 1980, Seasonal variation of 10-square mile probable maximum precipitation estimates, United States east of the 105th meridian: Hydrometeorological Rept. No. 53, NOAA, National Weather Service, Silver Spring, Md.

Kilpatrick, M.C., 1981, WATSTORE: A water data storage and retrieval system: U.S. Geol. Survey General Interest Publication, 15 p.

NOAA, 1973, Precipitation-frequency atlas of the western United States, Vols. 1-11: U.S. National Weather Service, Silver Spring, Md.

Schreiner, L.C. and Reidel, J.T., 1978, Probable maximum precipitation estimates, United States east of the 105th meridian: Hydrometeorological Report No. 51, NOAA, National Weather Service, Silver Spring, Md.

Solley, W.B., Chase, E.B., and Mann, W.B., IV, 1983, Estimated use of water in the United States in 1980: U.S. Geol. Survey Circ. 1001, 56 p.

Todd, D.K., 1970, The water encyclopedia: Port Washington, N.Y., Water Information Center, 559 p.

USGS, 1970, The national atlas of the United States of America: U.S. Dept. of Interior, Geological Survey.

USGS, 1981, A Guide to obtaining information from the USGS 1981: U.S. Geol Survey Circ. 777, 42 p.

USGS, 1983, National water summary 1983 - Hydrologic events and issues: U.S. Geol. Survey Water-Supply Paper 2250, 243 p.

U.S. Weather Bureau, 1961, Rainfall frequency atlas of the United States for durations from 30 minutes to 24 hours and return periods from 1 to 100 years: U.S. Weather Bureau Tech. Paper No. 40.

WRC, 1978, The nation's water resources, 1975-2000, second national assessment: U.S. Water Resources Council, Washington, D.C.

INDEX